The Sewing Machine Quick Guide

Paperback Edition ISBN 978-0-9900227-3-2

Published 2013 by Blodget Publishing, LLC.

blodgetpublishing@gmail.com

Dedication

This book is dedicated to my parents whose support I will always be grateful for.

Table of Contents

Introduction

The Sewing Machine Quick Guide will get you up to speed quickly on sewing machine basics. More than just a quick getting started book, complete information is presented so you will have a trusty reference guide long after you have mastered the basics. Step by step instructions show you how to use a sewing machine. Special attention is given to subjects like how to sew difficult fabrics and how to adjust the tension. Chapters on fabric, thread, stitches, needles and feet will help unlock your creativity.

Most sewing books have one or two pages about sewing machines because they are about sewing projects and not about sewing machines. This is the first modern and complete book dedicated to sewing machine basics! Both vintage machines and the latest electronic machines are covered. Quilting machines and embroidery machines have their own chapters.

If you are looking for an in-depth master guide to sewing machines please check out our other book "The Sewing Machine Master Guide". The Master Guide features over 100 types of home and industrial machines. The Master Guide also includes in-depth chapters on troubleshooting, adjustment and repair. For sergers check out our companion book "The Serger & Overlock Master Guide".

About this Book - The Sewing Machine Quick Guide was written as both an eBook and print book using optimized reflowable formatting for a perfect presentation on small or large devices. Everything from a basic eBook reader or Kindle to a large screen PC or Mac is supported.

Why the low price? Electronic publishing and print-on-demand is used with distribution to more than 80 countries worldwide. The pricing reflects this new technology and distribution model.

About the Author - Cliff Blodget learned to sew as a kid repairing tents and clothing. He became interested in sewing machines and discovered that with some adjustment, lubrication and TLC almost any sewing machine could be fine-tuned to sew like a new one. He recalls "Sewing machines have always fascinated me, they have a certain elegance and nostalgia factor. My favorites are the art deco style machines from the 1930's through the 50's".

Cliff has a BS degree in computer science and worked as an R&D Engineer and Technology Officer in electronics manufacturing. He never lost his interest in sewing machines. Later he became involved with sewing machines again, modifying and designing sewing equipment and processes for the production of sewn products such as cell phone cases and sportswear. He was frequently asked to recommend a good book on sewing machines, but could not find one that was up to date and comprehensive so he decided to write this book.

Sewing Machine Basics

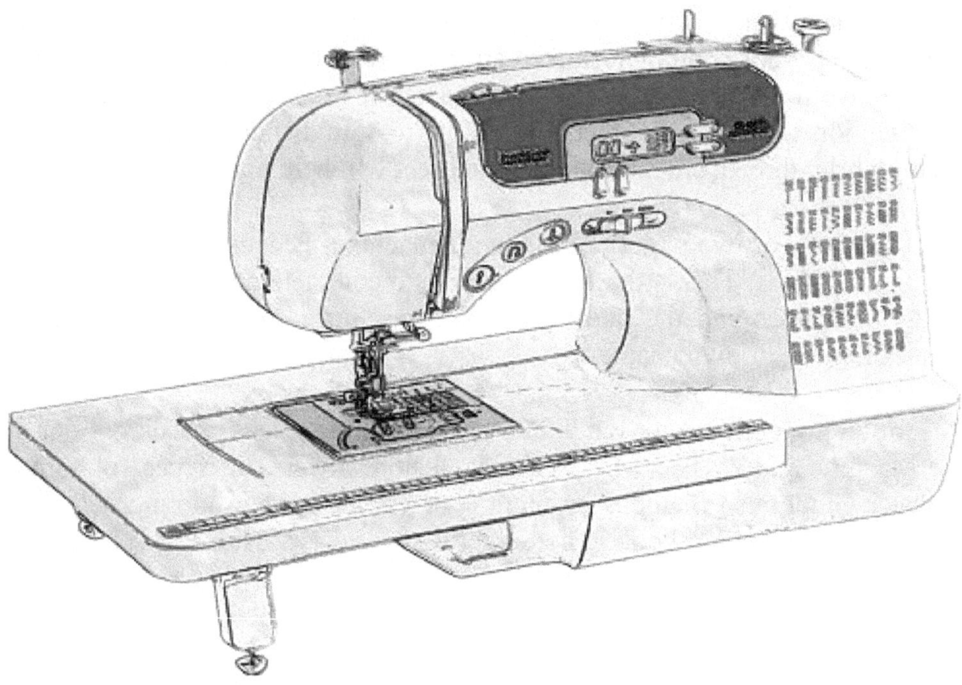

Some of the words that are used in this book are specific to sewing machines such "bobbin-case". If you encounter unfamiliar words there is a glossary at the end of the book.

Stitches

The primary job of a sewing machine is to make stitches. There are many types of stitches. Some types of stitches can only be made by hand while other types can be made by a sewing machine. The stitch type used for most sewing machines is the lockstitch. The lockstitch is versatile and does not come apart easily. The basic lockstitch can be made as a straight stitch, zigzag stitch or decorative stitch.

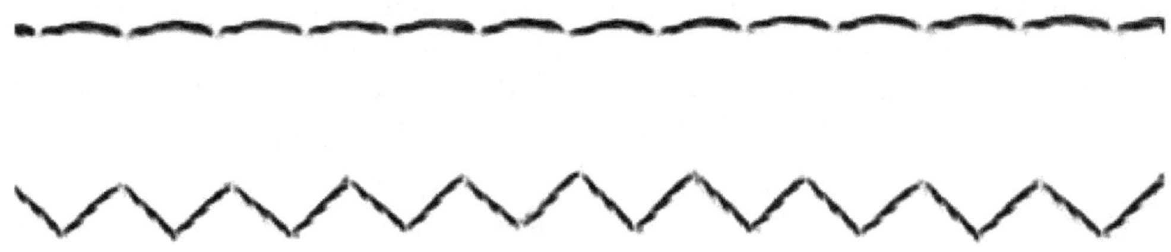

Straight stitch and zigzag stitch

- The straight stitch is primarily used for joining fabric and is the stitch used most often. The straight stitch can also be used to add extra stitching for decoration.
- The zigzag stitch is primarily used for joining fabrics that stretch because it can stretch better than a straight stitch. The zigzag stitch can also be used for decoration.
- Decorative stitches are used primarily for decoration but can be used for joining fabric.

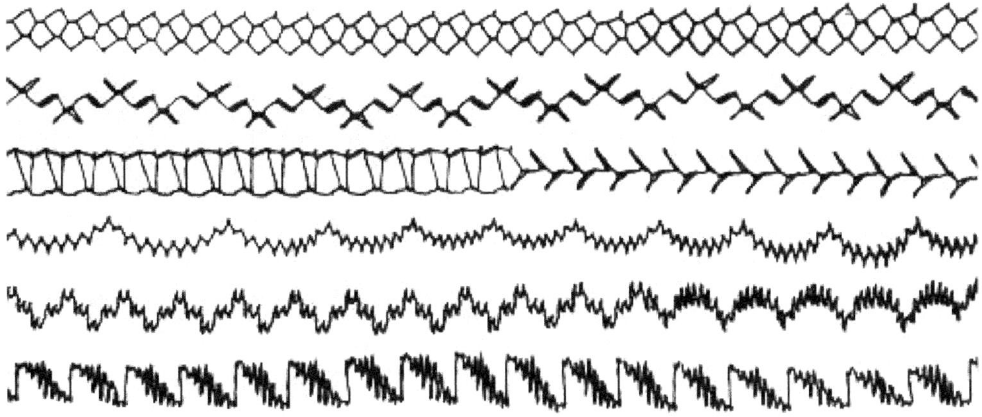

Decorative Lockstitch

A sewing machine makes zigzag stitches and decorative stitches by moving the needle from side to side as the stitches are being made. Some sewing machines can only sew a straight stitch, but most modern sewing machines can make the straight stitch, the zigzag stitch and decorative stitches.

A seam is a line of stitches sewn in succession. Seams that are used to attach or bind layers of fabric are known as construction seams.

For detailed information about stitches see the chapter "Stitches" later in the book.

Machine types

Home sewing machines come in models ranging form basic machines costing under $100 to expensive machines with many features that cost over $2000. Most sewing can be done with a basic machine like the one shown in the following picture.

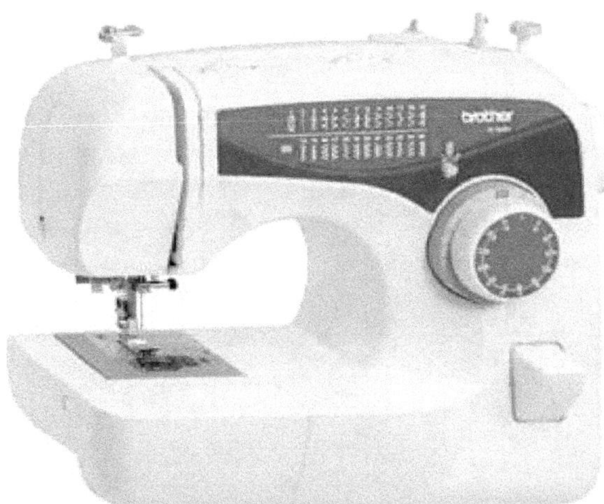

Brother XL2600i

Sewing machines come in home and industrial types. Home machines are compact and portable. Most modern home machines weigh under 25 pounds. Industrial sewing machines are made for continuous use in factories and can weigh more than 200 pounds! If you are interested in industrial machines they are covered in our book "The Sewing Machine Master Guide".

Heavy-duty home sewing machines are stronger versions of regular home machines. All cast iron machines built before 1970 are heavy duty (like the machine on the right side in the picture below). Cast iron machines are heavy, usually over 30 pounds.

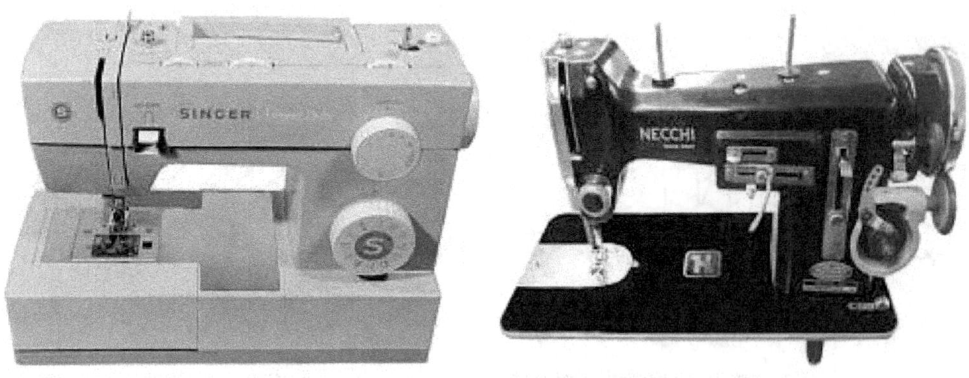

Modern Heavy Duty *Vintage Heavy Duty*

Mechanical machines

Mechanical machines are like a bicycle or a pair of scissors in that they are made only from metal and plastic parts and do not have electronics inside. Both machines in the picture above are mechanical machines. Older all-metal mechanical machines like the machine on the right can last over 100 years. Mechanical machines are still being made today by some manufacturers. People like mechanical machines because they are very reliable.

Electronic machines

In the 1970's electronic machines appeared on the market and have become increasingly more popular as time goes on. Electronic machines build on the basic mechanical machine and add electronics to control various aspects of the machine. Electronics enable advanced features. Some electronic machines have automatic thread cutters, automatic needle up/down, fancy decorative stitch patterns and more. The machine in the following picture is an electronic machine.

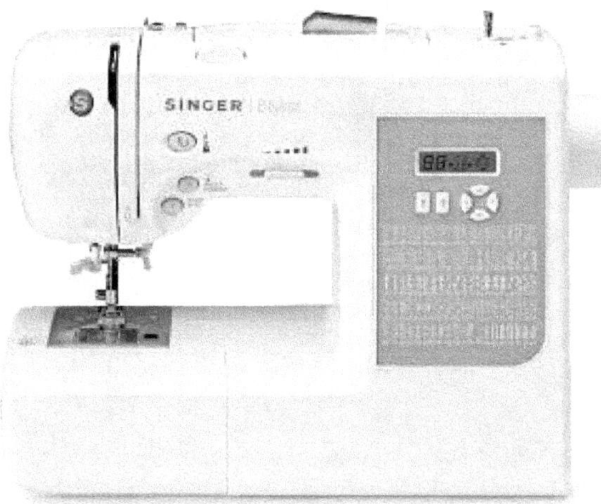

Singer Stylist 7258

If a machine has any of the following features it is an electronic machine: needle up/down button, automatic thread cutter, auto back tacking, touch screen or LCD display, cursor controls or arrow keys, programmable stitches, memory, fonts, monograms, USB data connection, etc.

Sergers

Sergers are special machines that only sew on the edge of the fabric. As you can see in the next picture a serger doesn't even look like a regular sewing machine. Sergers make a type of stitch called the overlock stitch. The overlock stitch is used for edge finishing. Edge finishing is used to secure the edge of the fabric to prevent the fabric from fraying. The overlock stitch can also be used to join the edges of two pieces of fabric at the same time as securing the edges from fraying.

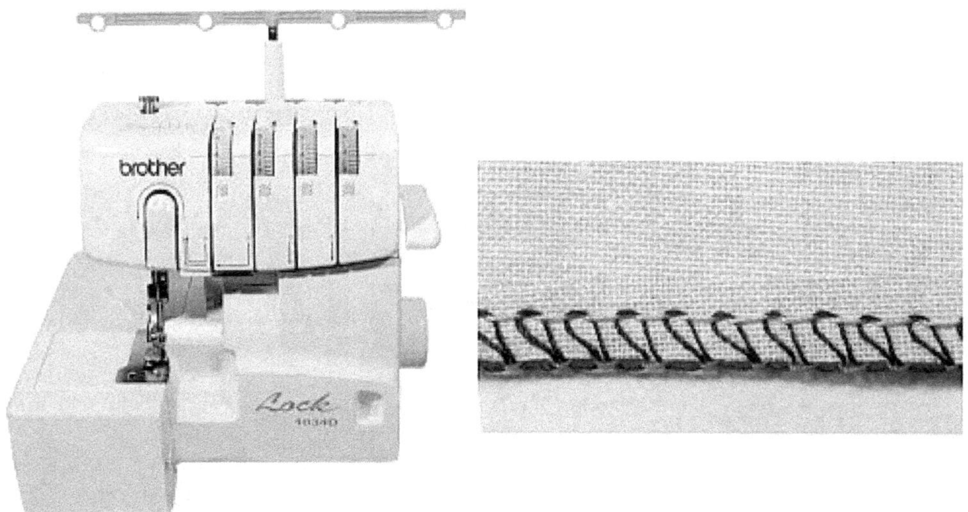

Home Serger ***Overlock Stitch***

Although a zigzag sewing machine can make a stitch called an overedge stitch that will work for occasional edge finishing, a serger is preferred if you are doing a lot of edge finishing. Sergers are much faster and make a true overlock stitch that is better looking and stronger than the overedge stitch. If you are interested in sergers check out our companion books "The Serger & Overlock Quick Guide" and "The Serger & Overlock Master Guide".

New Machines

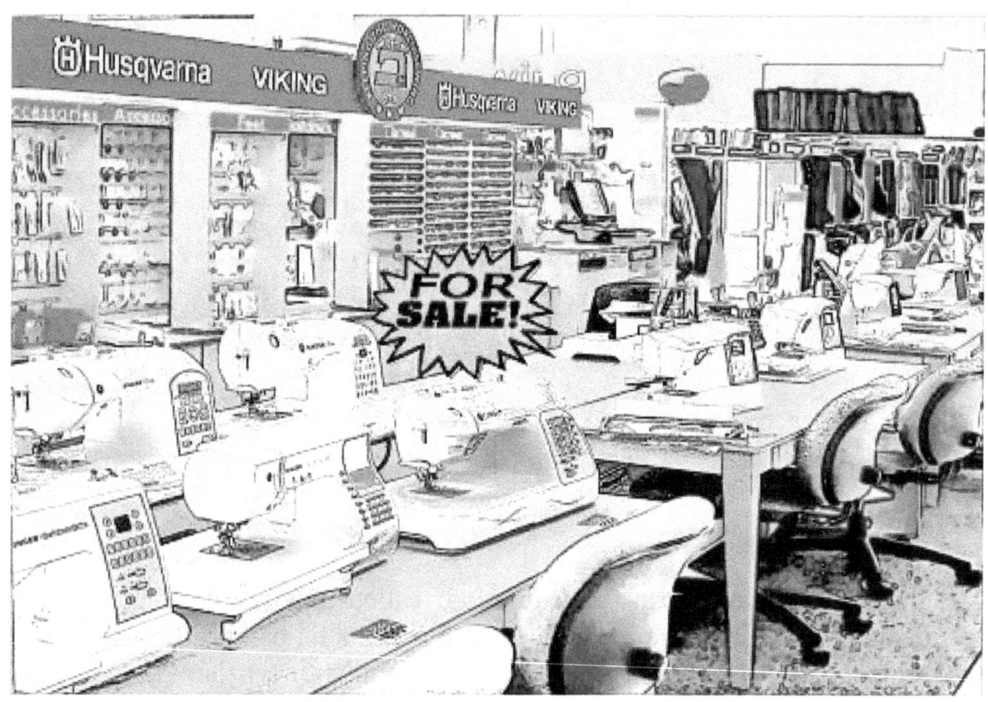

This chapter gives information and buying pointers for new machines and then lists recommended machines in several categories. These are my personal recommendations and opinions. Compare them with your requirements and decide what will work for you. The machines presented in this chapter were selected for the following reasons:

- Best quality and features for the cost.
- Models that are widely available in most markets.
- Models with stand-out performance or features.

All of the major brands (like Brother, Singer, Janome, Juki, Pfaff, Viking, Bernina, Elna, Baby Lock, etc) make good machines. In some countries there are also other great brands. The machines listed in this chapter are examples of machines that are a good value at the present time, but keep in mind there are other models that may be good and also keep in mind that new models are frequently being introduced. Use the machines listed in this book as a basis for comparison. Check to see what you are able to get in your location. Prices change as well, when you are ready to buy a machine check the latest prices and check to see if any of the machines you like are on sale.

I have no connection or affiliation with the brands that manufacture or sell the models that appear in this book, the models are listed because I feel they represent the best quality for the cost. The recommended models are 120 volt machines sold in the United States. Similar machines with different model numbers and voltages (such as 240V) are sold in other countries. The prices that are listed were at the time of writing of the book and may change from time to time.

For general sewing - For general sewing any of the machines listed in this chapter will work well. Even the inexpensive machines use standard feet and with the correct foot will do any type of operation that you may need. There is no difference in the stitch quality between inexpensive machines and premium machines. Higher cost machines are often marketed as being of superior quality, but this is usually not the case. The biggest difference is that expensive machines tend to have more features. Features can include decorative stitch types, needle threaders, free arms, buttonholers, etc.

For heavy sewing - If you mainly sew heavy materials such as jeans, jackets or camping gear, then get a heavy duty machine or a high speed straight stitch machine. For really heavy materials look into an industrial machine.

Machines are similar! - Most sewing machines are actually more similar in performance and function then you would think. For example a basic machine from Brother is probably very similar to a basic machine in the same price range from Janome or Singer. Pick the machine that is the best fit for you and your type of sewing and be happy!

Retail sales - Unfortunately some retail sales people have little sewing knowledge or don't have any sewing experience at all. Most stores put out miss-information about inexpensive machines, telling horror stories about quality problems and bad service. This is meant to scare customers into buying more expensive machines.

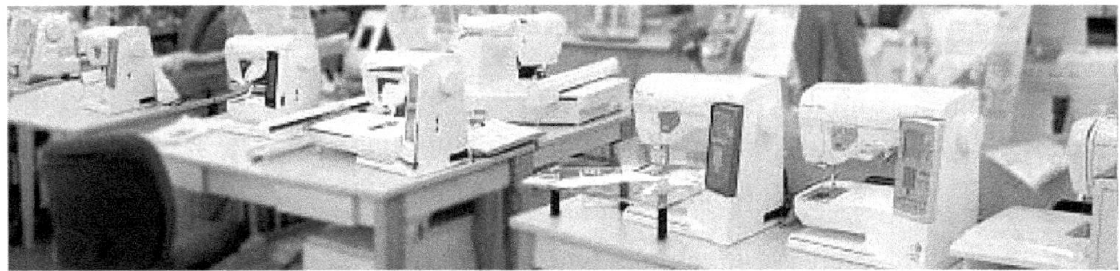

Information - Download and read the owners manuals to learn about specific machines you are interested in. Test machines yourself if possible to develop your own opinion. The best sources of information are Internet forums, sewing groups, books, and experienced people you know.

Simple is good! - If you are not sure what type of machine is best for you, get a basic machine and save yourself some money. The stitch quality of all machines is about the same if they are adjusted properly. A machine with automatic features won't make up for a lack of knowledge and skill. Many people get fancy machines because they are worried that a simple machine may not do what they need. Actually most types of sewing can be done with a very basic machine. A basic machine is easier to learn, easier to adjust, easier to troubleshoot and may be more reliable!

Stitch selection - A large selection of stitches is not important for most sewing. Ninety five percent of all sewing can be done with two basic stitch types, the straight stitch and the zigzag stitch. Very little sewing is actually done with fancy decorative stitches like the ones in the picture below. If you do like decorative stitches then by all means you should get a machine with a selection of decorative stitches, my point is that for most sewing it is not mandatory to have decorative stitches at all.

Adjustment - The correct needle, thread and adjustment of a machine makes a huge difference, more difference then the brand and model of machine. An improperly adjusted $2000 machine will sew considerably worse than a properly adjusted $89 machine.

Automatic features - Automatic features are things like needle threaders, automatic thread cutters, needle up/down and auto back tacking. These features are described in the feature list at the end of this chapter. Features are for convenience. In general features make no difference as far as what you can sew or the quality of your sewing. On the other hand some of these features may be beneficial for some types of sewing. For example in quilting an automatic thread cutter may be good to have because with a large quilt on your machine it can be hard to cut the bottom thread at the end of a seam. You should make an informed choice as to what features you really need.

Speed - Sewing machine speed is measured in stitches per minute (SPM). Most home machines run at a speed of 750 to 1100 SPM, this if fine for most uses. The high speed straight stitch machines sew at 1500 or 1600 SPM and the added speed comes in handy if you are doing a lot of sewing of larger items. You will not be able to tell much speed difference between a machine that can sew at 800 SPM and one that can sew at 900 SPM, but a machine that can sew 1100 SPM will feel somewhat faster. The high speed straight stitch machines that can run at 1500 or 1600 SPM do feel noticeably faster and are nice if you are doing a lot of sewing of large items on a daily basis.

Needle size - All machines can take needles up to a size 16 needle, this is fine for most sewing including sewing jeans and blankets. Some machines can take a size 18 needle, you may want to use a size 18 needle with very thick thread for sewing heavy items that are giving you trouble with a size 16 needle.

Feet - More then features, a selection of feet will enable a machine to do many kinds of operations. All machines shown in this chapter take standard low shank feet except the high speed straight stitch machines which take high shank feet. There are many types of accessory feet available, see the chapter on Presser Feet for an overview of feet types and to learn about the shank types.

Inexpensive basic machines

Basic inexpensive machines from Brother or Singer priced at $100 or less sew better than you would expect, in fact the stitch quality is just as good as an expensive machine. The inexpensive machines have limited features, but some sewers that want a simple machine consider this an advantage.

Some first time sewers buy inexpensive machines and have problems, but most of their problems are due to operator error while learning and would happen with any machine. Some frustrated users give these machines bad reviews on-line, they assume that because the machine is inexpensive their problems must be due to the machine. This explains most of the bad reviews of inexpensive machines you will see.

Brother LS-2125i

Description – This is a basic mechanical machine ($75 Amazon.com). It has a single control knob that controls stitch length, zigzag width, stitch pattern and needle position, this is good if you want simplicity, but not good if you want to adjust these settings individually. For general purpose sewing it is not a problem. The front facing bobbin-case pops out of the machine for easy cleaning.

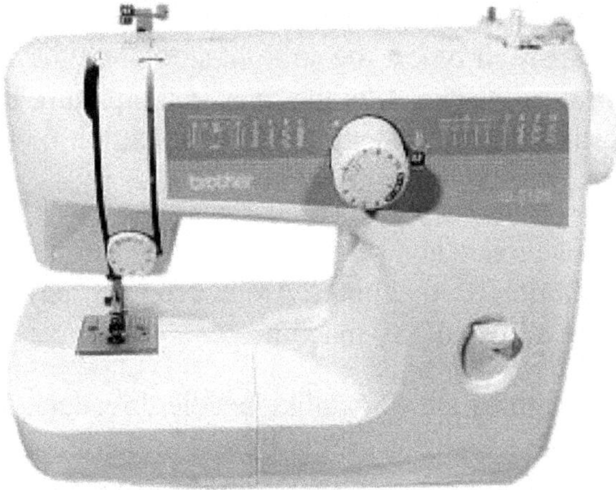

Brother LS2125

Features – Machine weight is about 13 pounds, built in handle for easy carrying, needle threads from front to back, Twin needle capable, straight stitch, zigzag, basic utility stitches, Needle sizes up to a size 16-(100), buttonholer, free arm, comes with several feet, darning and free motion quilting is possible with the included feed dog cover plate (feed dogs do not drop), non-adjustable presser foot pressure, no needle threader, maximum speed is 900 SPM.

Brother XL-2600i

Description – This is a basic mechanical machine ($85 Amazon.com). Has top-loading bobbin. The needle plate removes with two screws and then the bobbin-case can lift out for cleaning.

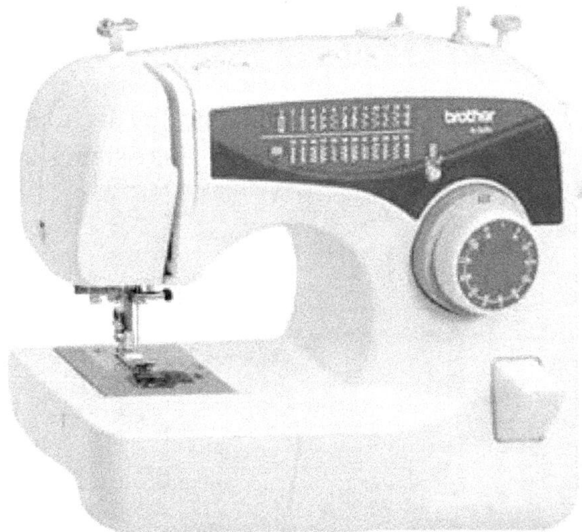

Brother XL2600i

Features – Has a stitch pattern control knob and a separate stitch length dial and stitch width controls. Machine weight is about 11 pounds, built in handle for easy carrying, needle threads from front to back, Twin needle capable, straight stitch, zigzag, basic utility stitches and a selection of decorative stitches, Needle sizes up to a size 16-(100), auto-sizing buttonholer, free arm, comes with several feet, darning and free motion quilting is possible with the included feed dog cover plate (feed dogs do not drop), non-adjustable presser foot pressure, needle threader, maximum speed is 800 SPM.

Singer Tradition 2259

Description – This is a basic mechanical machine ($78 Amazon.com). The front facing bobbin-case pops out for easy cleaning.

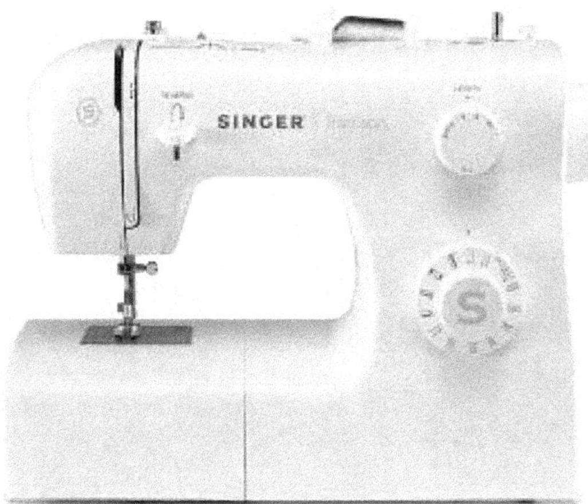

Singer Tradition 2259

Features – Has a stitch pattern control knob and a separate stitch length knob and stitch width dial. Machine weight is about 13 pounds, built in handle for easy carrying, needle threads from front to back, Twin needle capable, straight stitch, zigzag, basic utility stitches and a selection of decorative stitches, Needle sizes up to a size 18-(110), automatic buttonholer, free arm, comes with several feet, darning and free motion quilting is possible with the included feed dog cover plate (feed dogs do not drop), non-adjustable presser foot pressure, no needle threader, maximum speed is 800 SPM.

Heavy duty machines

Heavy-duty machines weigh more, are stronger, run smoother and are more abuse tolerant than inexpensive machines, but are otherwise similar. The Singer models are a fantastic value at the prices listed.

Singer 4411

Description – Basic heavy-duty machine ($99 Amazon.com). This machine is the same as the 4423 featured next, but has no needle threader, buttonholer is not auto-sizing and has fewer stitch types. This machine is fast (Maximum speed of 1100 SPM).

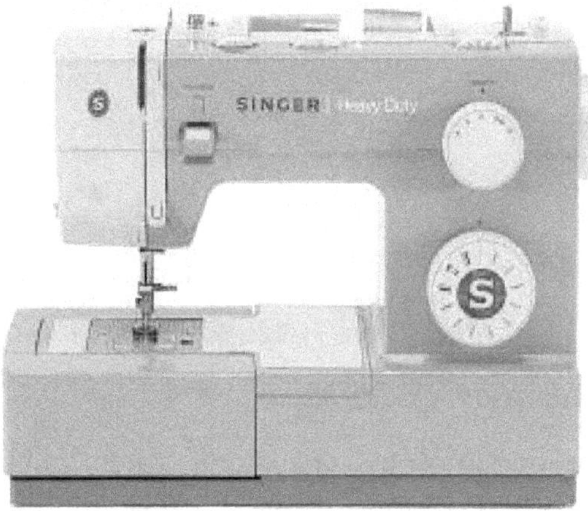

Singer 4411

Features – There are separate controls for stitch pattern, stitch length, stitch width and needle position. Top-loading drop-in bobbin, machine weight is about 13 pounds, built in handle for easy carrying, needle threads from front to back, Twin needle capable, straight stitch, zigzag, basic utility stitches and a selection of decorative stitches, Needle sizes up to a size 18-(110), buttonholer, free arm, comes with several feet, feed dogs drop for darning and free motion quilting, adjustable presser foot pressure, no needle threader.

Singer 4423

Description – Basic heavy-duty machine ($119 Amazon.com). This machine is fast (Maximum speed of 1100 SPM).

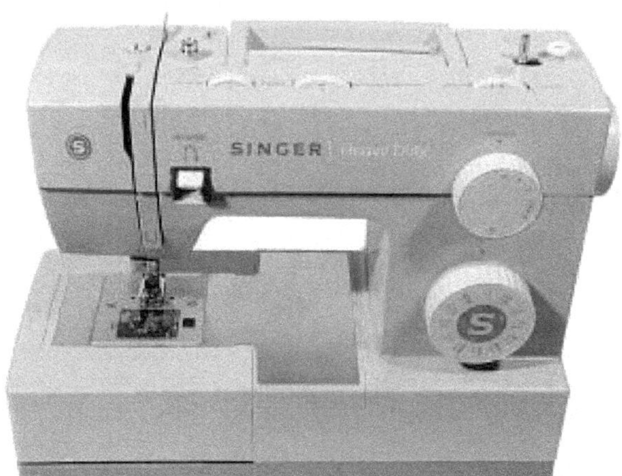

Singer 4423

Features – There are separate controls for stitch pattern, stitch length, stitch width and needle position. Top-loading drop-in bobbin, Machine weight is about 13 pounds, built in handle for easy carrying, needle threads from front to back, Twin needle capable, straight stitch, zigzag, basic utility stitches and a selection of decorative stitches, Needle sizes up to a size 18-(110), auto-sizing buttonholer, free arm, comes with several feet, feed dogs drop for darning and free motion quilting, adjustable presser foot pressure, needle threader.

Janome HD1000

Description – Basic heavy-duty machine ($299 Amazon). Front facing bobbin, bobbin-case pops out for easy cleaning. Body is cast aluminum.

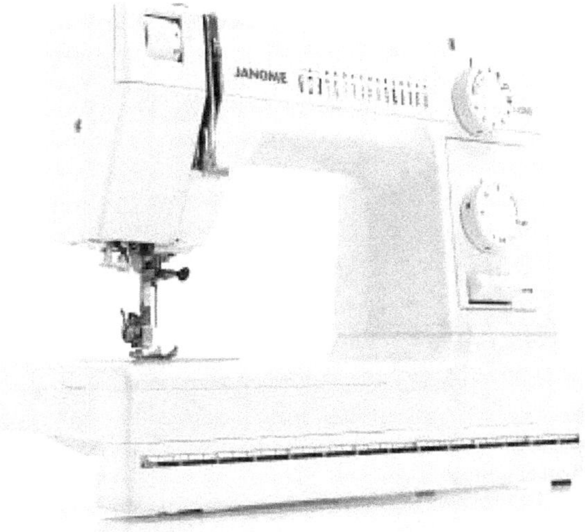

Janome HD-1000

Features – Equipped with stitch pattern control knob including stitch width settings and a separate stitch length dial. Machine weight is about 16 pounds, built in handle for easy carrying, needle threads from front to back, twin needle capable, straight stitch, zigzag, basic utility stitches and a selection of decorative stitches, needle sizes up to a size 16-(100), buttonholer, free arm, comes with several feet, feed dogs drop for darning and free motion quilting, adjustable presser foot pressure, needle threader, maximum speed is 860 SPM.

Janome HD3000

Description – Basic heavy-duty mechanical machine ($349 Amazon). The needle plate removes with one screw and then the top loading bobbin-case can lift out for easy cleaning.

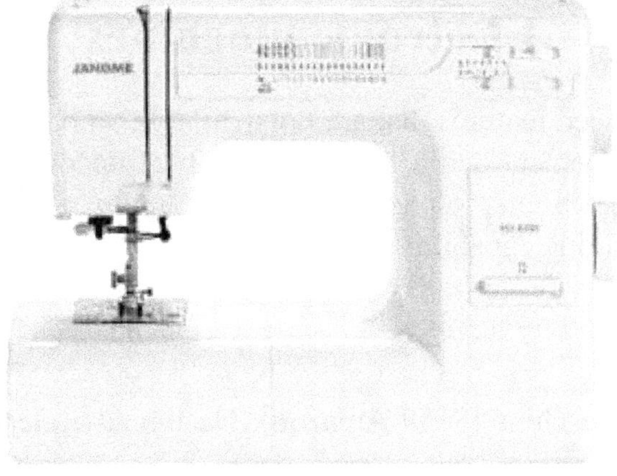

Janome HD-3000

Features – Equipped with stitch pattern control knob and a separate stitch length dial and stitch width dial. Machine weight is about 18 pounds, built in handle for easy carrying, needle threads from front to back, twin needle capable, straight stitch, zigzag, basic utility stitches and a selection of decorative stitches, needle sizes up to a size 16-(100), auto-sizing buttonholer, free arm, comes with several feet, feed dogs drop for darning and free motion quilting, adjustable presser foot pressure, needle threader, maximum speed is 860 SPM.

Janome S-750 and S-950

Description – These are flat bed machines that were designed for school use, so they are quite heavy-duty. They are among the few flat bed machines still being made that will mount in a standard sewing table. Cost for these machines on-line is about $250-$350

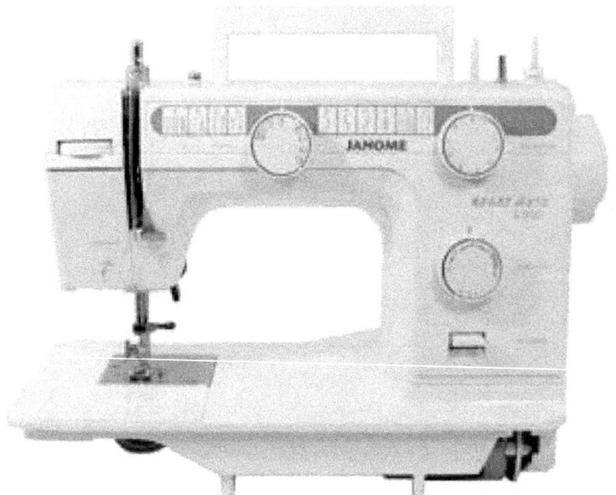

Janome Classmate S950

S-950 - has a front-facing bobbin.

S-750 - has a top-loading drop-in bobbin.

Features – Equipped with stitch pattern control knob and a separate stitch length dial and stitch width dial. Machine weight is about 18 pounds, built in handle for easy carrying, needle threads from front to back, twin needle capable, straight stitch, zigzag, basic utility stitches and a selection of decorative stitches, needle sizes up to a size 16-(100), auto-sizing buttonholer, flat bed (no free arm), comes with several feet, feed dogs drop for darning and free motion quilting, adjustable presser foot pressure, needle threader, maximum speed is 860 SPM.

Electronic machines

Electronic machines have advanced features that are not available on basic and heavy duty machines such as needle up/down, auto reinforcement and a larger selection of decorative stitches. Some models have automatic thread cutters, knee lifters, alphanumeric monogram stitches, computer interface, user programmable stitches, stitch editing capability, etc.

Brother CS-6000i

Description – Basic electronic machine ($159 Amazon). No thread cutter, comes with good selection of feet and accessories.

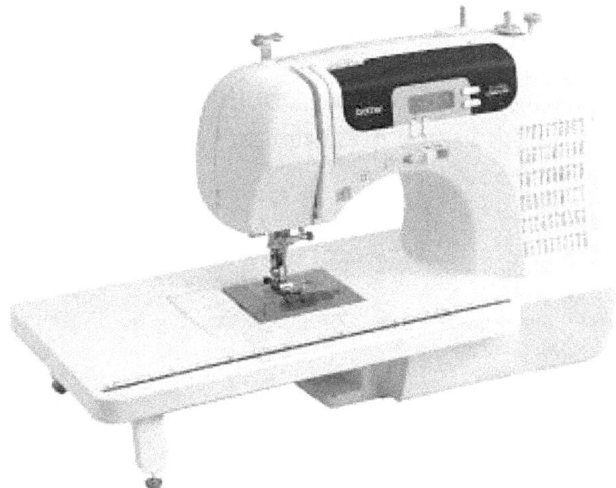

Brother CS-6000i

Features – There are separate controls for stitch pattern, stitch length, stitch width. Needle position is adjustable. Top-loading drop-in bobbin, machine weight is about 13 pounds, built in handle for easy carrying, needle threads from front to back, twin needle capable, straight stitch, zigzag, basic utility stitches and a large selection of decorative stitches, needle sizes up to a size 16-(100), auto-sizing buttonholer, free arm, comes with several feet, feed dogs drop for darning and free motion quilting, non-adjustable presser foot pressure, needle threader, maximum speed is 850 SPM, LCD display, speed control knob, start-stop button, needle up-down button, reverse/reinforcement button.

Singer 7258 Stylist

Description – Basic electronic machine($164 Amazon). No thread cutter, comes with good selection of feet and accessories.

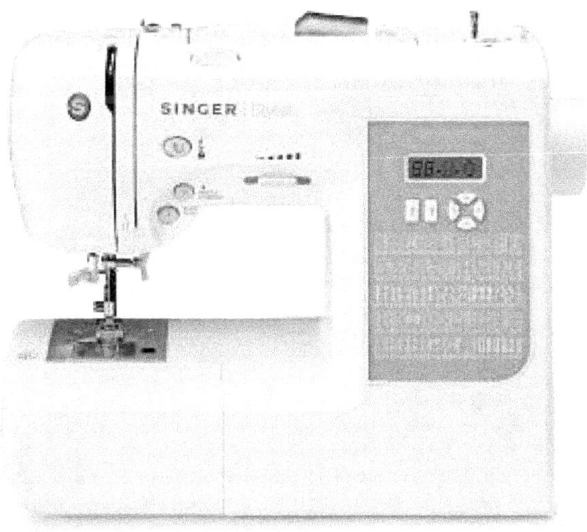

Singer Stylist 7258

Features – There are separate controls for stitch pattern, stitch length, stitch width. Needle position is adjustable. Top-loading drop-in bobbin, machine weight is about 15 pounds, built in handle for easy carrying, needle threads from front to back, twin needle capable, straight stitch, zigzag, basic utility stitches and a large selection of decorative stitches, needle sizes up to a size 18-(110), auto-sizing buttonholer, free arm, comes with several feet, feed dogs drop for darning and free motion quilting, non-adjustable presser foot pressure, needle threader, maximum speed is 750 SPM, LCD display, speed control knob, start-stop button, needle up-down button, reverse/reinforcement button. Needle plate removes with 2 screws then bobbin-case is removable.

Brother SE400

Description - Electronic machine with embroidery capability ($358 Amazon). If you don't do embroidery then get the PC-420 or Singer 9960 in the same price range.

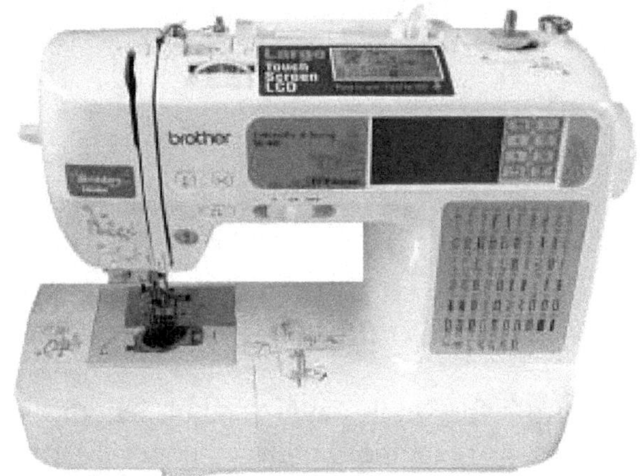

Brother SE400

Features – There are separate controls for stitch pattern, stitch length and stitch width. Needle position is adjustable. Top-loading drop-in bobbin, machine weight is about 11 pounds, built in handle for easy carrying, needle threads from front to back, twin needle capable, straight stitch, zigzag, basic utility stitches and a large selection of decorative stitches, needle sizes up to a size 16-(100), auto-sizing buttonholer, free arm, comes with several feet, feed dogs drop for darning and free motion quilting, non-adjustable presser foot pressure, needle threader, maximum speed is 710 SPM, LCD display, speed control knob, start-stop button, needle up-down button, reverse/reinforcement button, easy to remove bobbin-case.

Special Features - This machine has a thread cutter and also does embroidering (up to 4"x4"). Has 5 embroidery letter fonts and does monograms. Includes a USB computer interface. Uses industry standard .pes file format.

Singer 9960 Quantum Stylist

Description - Feature rich electronic machine ($329 Amazon) – Good value for an electronic machine with thread cutter.

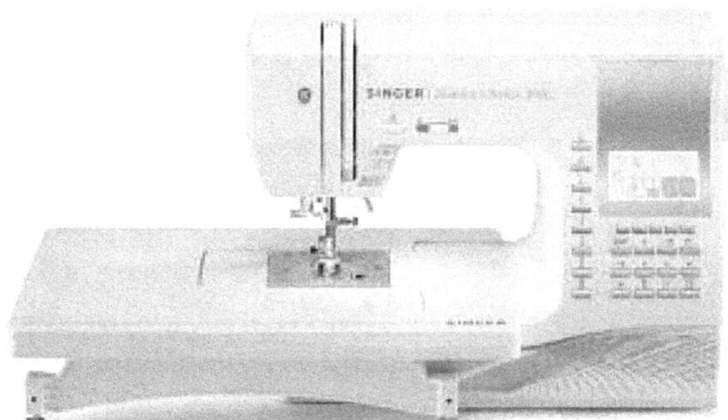

Singer Quantum Stylist 9960

Features – There are separate controls for stitch pattern, stitch length, stitch width. Needle position is adjustable. Top-loading drop-in bobbin, machine weight is about 15 pounds, built in handle for easy carrying, needle threads from front to back, twin needle capable, straight stitch, zigzag, basic utility stitches and a large selection of decorative stitches, needle sizes up to a size 16-(100), auto-sizing buttonholer, free arm, comes with several feet, feed dogs drop for darning and free motion quilting, adjustable presser foot pressure, needle threader, horizontal thread spool, maximum speed is 850 SPM, LCD display, speed control knob, start-stop button, needle up-down button, reverse/reinforcement button.

Special Features - Auto thread cutter, alphanumeric monogram stitches, combined pattern stitches, includes a large selection of feet. Needle plate removes with 2 screws then bobbin-case is removable.

Brother PC-420 PRW

Description – Feature rich electronic machine ($409 Amazon.com). This machine seems stronger than other electronic machines in this price range and is the lowest cost machine available with advanced features like a knee lifter and user programmable stitches.

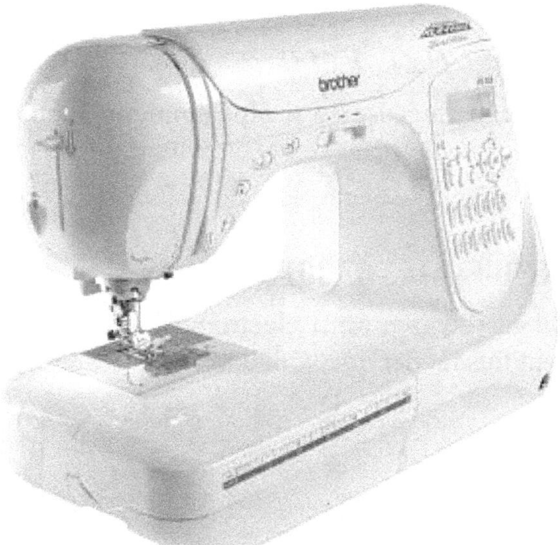

Brother PC-420

Features – There are separate controls for stitch pattern, stitch length, stitch width. Needle position is adjustable. Top-loading drop-in bobbin, machine weight is about 17 pounds, built in handle for easy carrying, needle threads from front to back, twin needle capable, straight stitch, zigzag, basic utility stitches and a large selection of decorative stitches, needle sizes up to a size 16-(100), auto-sizing buttonholer, free arm, comes with several feet, feed dogs drop for darning and free motion quilting, adjustable presser foot pressure, needle threader, maximum speed is 850 SPM, LCD display, speed control knob, start-stop button, needle up-down button, reverse/reinforcement button.

Special Features - Knee lifter, auto thread cutter, alphanumeric monogram stitches, user programmable stitches, combined pattern stitches, includes a large selection of feet, easy to remove bobbin-case.

Juki HZL-F600 (Exceed series)

Description – Feature rich and powerful electronic machine ($849 eBay.com).

Features – There are separate controls for stitch pattern, stitch length, stitch width. Needle position is adjustable. Top-loading drop-in bobbin, machine weight is about 22 pounds, built in handle for easy carrying, needle threads from front to back, twin needle capable, straight stitch, zigzag, basic utility stitches and a large selection of decorative stitches, needle sizes up to a size 16-(100), auto-sizing electronic buttonholer, free arm, comes with several feet, feed dogs drop for darning and free motion quilting, adjustable presser foot pressure, needle threader, horizontal thread spool, maximum speed is 900 SPM, LCD display, speed control knob, start-stop button, needle up-down button, reverse/reinforcement button.

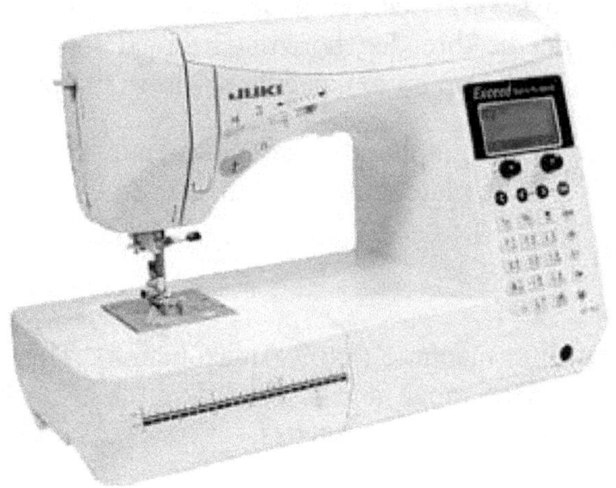

Juki HZL-F600

Special Features - Knee lifter, auto thread cutter, alphanumeric monogram stitches, includes a large selection of feet, auto bar-tacking, needle plate removes with 2 screws then bobbin-case is removable, professional electronic buttonholer.

Janome Memory Craft 6300 (MC-6300P)

Description - Heavy-duty flat bed (no free arm) electronic machine ($1199 Amazon). Janome makes other more expensive machines in this model line with additional features.

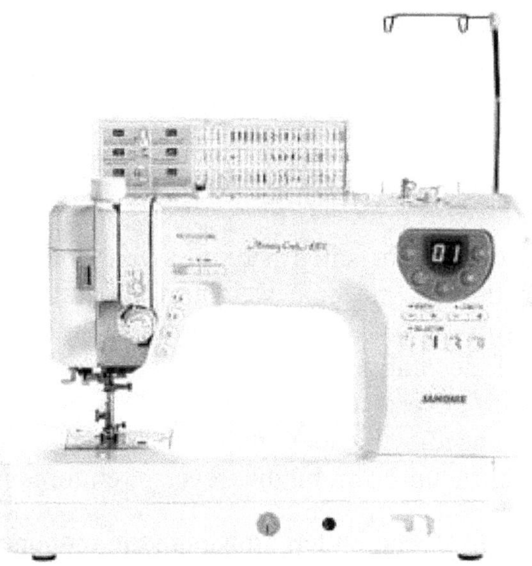

Janome Memory Craft 6300P

Features – There are separate controls for stitch pattern, stitch length, stitch width. Needle position is adjustable. Top-loading drop-in bobbin, machine weight is about 26 pounds, built in handle for easy carrying, needle threads from front to back, twin needle capable, straight stitch, zigzag, basic utility stitches and a large selection of decorative stitches, needle sizes up to a size 18-(110), auto-sizing buttonholer, flat bed (no free arm), comes with several feet, feed dogs drop for darning and free motion quilting, adjustable presser foot pressure, needle threader, maximum speed is 1000 SPM, LCD display, speed control knob, start-stop button, needle up-down button, reverse/reinforcement button.

Special Features - Knee lifter, auto thread cutter, built in thread tower, stitch editing capability, combined pattern stitches, includes a large selection of feet, auto bar-tacking, needle plate removes with 2 screws then bobbin-case is removable, long bed with 9 inches of harp space, separate bobbin winding motor.

High speed straight stitch machines

These models sew straight stitches only. They are powerful, fast and very smooth. They have excellent low speed control and can sew through thick materials with ease. These are flat bed machines with extended beds and harp space (no free arm). Harp space is the room for fabric to the right of the needle. Machines with a lot of harp space are nice if you are sewing large items like quilts, blankets and jackets. Industrial L style bobbins are used in these machines. If you do a lot of sewing and can afford one of these machines they are strongly recommended. Some people have one high speed straight stitch machine and a second lower cost zigzag machine for when they need zigzag and decorative stitches. Most people with two machines find that they do most of their sewing on the high speed straight stitch machine.

Brother PQ-1500s

Description - Heavy-duty electronic high speed straight stitch machine ($669 Amazon).

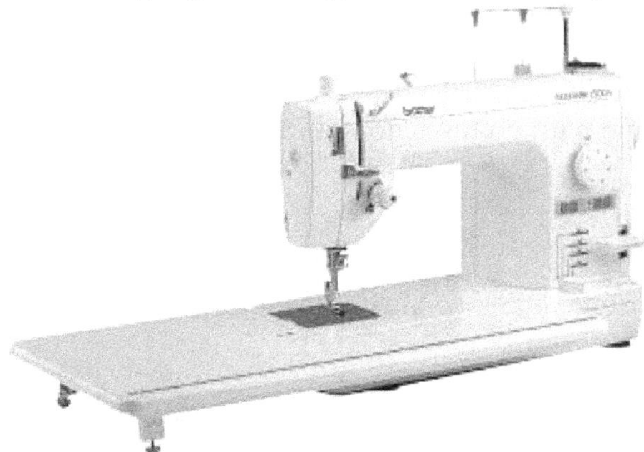

Brother PQ-1500S

Features – There are separate controls for stitch length and stitch width. High shank feet, side-loading bobbin, SA159 or L style bobbin, machine weight is about 24 pounds, built in handle for easy carrying, needle threads from left to right, straight stitch only, Needle sizes up to a size 18-(110), needle type HLX5, flat bed (no free arm), comes with several feet, feed dogs drop for darning and free motion quilting, adjustable presser foot pressure, needle threader, maximum speed is 1500 SPM, maximum stitch length 7mm, speed control knob, start-stop button, needle up-down button, reverse/reinforcement button.

Special Features - Knee lifter, auto thread cutter, built in thread tower, needle plate removes with 2 screws, bobbin-case is removable with no tools, long bed with 8.5 inches of harp space, pin feed for hard to feed fabrics. Pin feed is a type of feed mechanism that uses a pin that protrudes form the feed dogs through the fabric to ensure that multiple layers of fabric will feed in synchronization without slippage.

Janome 1600P-QC

Description - Heavy-duty electronic high speed straight stitch machine ($799 on-line stores).

Features – There are separate controls for stitch length and stitch width. High shank feet, side-loading bobbin, SA159 or L style bobbin, machine weight is about 26 pounds, built in handle for easy carrying, needle threads from left to right, straight stitch only, Needle sizes up to a size 18-(110), needle type HLX5, flat bed (no free arm), comes with several feet, feed dogs drop for darning and free motion quilting, adjustable presser foot pressure, needle threader, maximum speed is 1600 SPM, maximum stitch length 6mm, speed control knob, start-stop button, needle up-down button, reverse/reinforcement button.

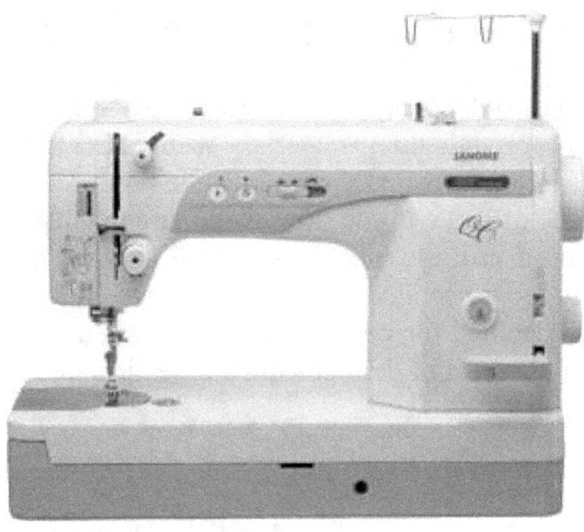

Janome 1600P-QC

Special Features - Knee lifter, auto thread cutter, built in thread tower, needle plate removes with 2 screws, bobbin-case is removable with no tools, long bed with 8.5 inches of harp space.

Juki TL-2010Q

Description - Heavy-duty electronic high speed straight stitch machine ($849 on-line stores).

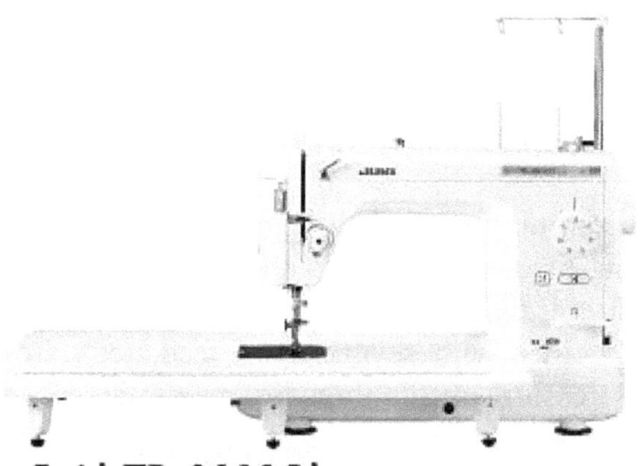

Juki TL-2000Qi

Features – There are separate controls for stitch length and stitch width. High shank feet, side-loading bobbin, SA159 or L style bobbin, machine weight is about 25 pounds, built in handle for easy carrying, needle threads from left to right, straight stitch only, Needle sizes up to a size 18-(110), needle type HLX5, flat bed (no free arm), comes with several feet, feed dogs drop for darning and free motion quilting, adjustable presser foot pressure, needle threader, vertical thread spool, maximum speed is 1500 SPM, maximum stitch length 6mm, speed control knob, start-stop button, needle up-down button, reverse/reinforcement button.

Special Features - Knee lifter, auto thread cutter, built in thread tower, needle plate removes with 2 screws, bobbin-case is removable with no tools, long bed with 8.5 inches of harp space, thread trimmer can be controlled from the foot pedal.

How to test machines and do research

First do your research on the Internet. Before you go to a sewing machine dealer and try out machines you should research the machines you are interested in and read user feedback and comments on Amazon and on forums. This is very important because it takes a long time to really know a sewing machine. During a quick test you may think that you really like a machine only to find that after a few months you don't really like it. This can be avoided by reading as much as you can about the machines you are interested in. If you find that a certain machine is overwhelmingly liked by users that do similar types of sewing then it is a good bet that you will like it as well. Or you may read about certain issues with a machine that needs further consideration.

Go on-line and download the owners manuals for the machines you are interested in. If the manuals are not available on-line then the manufacturer is hiding something and does not want you to do a direct comparison of their products. Reading the user manual can give you a lot of information about the machine that you will not find elsewhere. Retail sales people sometimes have no idea of what they are talking about, but the user manual will always tell you a lot.

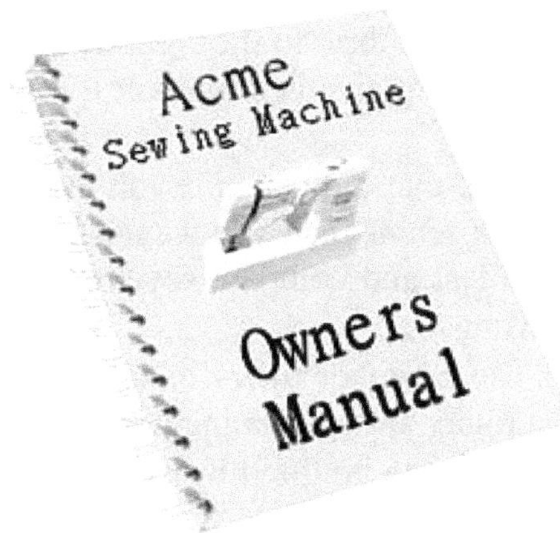

Be careful about getting opinions from users that have just bought a new machine and are trying to convince themselves that they made a good purchase but have not extensively used the machine yet. Ask them how long they have had the machine and what they have actually done with it. This includes on-line opinions.

Test sew the machines if you can. Bring the type of fabric and thread that you use for your projects. Do as much sewing on the machines as possible. Try out all of the features that you are most concerned with. Take notes if you are testing several machines. If you know what the most troublesome operations are for you (such as sewing the crotch seams of jeans or attaching zippers) then try those operations with the machines you are testing.

Feature checklist - basic and heavy duty machines

This is a general checklist of features and items that you may find are important for your type of sewing. See the glossary at the end of the book for a definition of any of these terms.

- **Adjustable feed dog height** - Allows you to adjust the feed dog height for different types and thicknesses of fabric.
- **Adjustable presser foot pressure** – Lets you adjust the force that the presser foot applies to the fabric, this is good to have if you are sewing very light weight or heavy weight fabrics.
- **Automatic needle threader** – This is a device that helps you thread the needle. Some people never use needle threaders, other people think that they are a good thing. Older machines and industrial machines do not have needle threaders. If you have shaky hands or poor manual dexterity you may really benefit from a needle threader.
- **Removable bobbin-case** - HA-1 bobbin cases and some top loading bobbin cases are removable.

A removable bobbin case has the advantage of being able to use several bobbin cases, with each case adjusted to a different thickness of thread. Removable bobbin cases also make the machine easier to troubleshoot and service.

- **Decorative stitches** – If you need them, make sure you get a machine that can make them, many older machines are straight stitch or zigzag only. Some older machines need stitch pattern cams to make decorative stitches while newer machines have built in cams or use electronic control circuits instead of cams.
- **Drop feed dogs** – This is a lever or control on some machines that drop the feed dogs for darning and free motion quilting. Some machines use a feed dog cover plate do the same thing.
- **Ease of access for maintenance and oiling** - Some machines have easy to access internal parts while others need to be almost completely disassembled to access anything.
- **Flat bed** – If you don't need a free arm then flat beds tend to be more stable.
- **Free arm** – Good to have if you are sewing some types of sleeves and cuffs.
- **Harp size** – This is the space for the fabric to the right of the needle. Most home machines have about 6-1/2 inches of harp space. Harp size is important if you are quilting, sewing tents or other bulky items.
- **Piercing power** - This is the force that the machine can drive the needle downward with. If you need to sew heavy fabrics or thick seams then you want a machine with high piercing power. The best way to determine this is to test and compare several models of machines with the kinds of fabrics or seams you will be sewing.
- **Hook type** – Home machines can be either rotary or oscillating type. Both types work well. See the glossary for information on rotary and oscillating hooks.
- **Knee lifter** – Allows the presser foot to be lifted by a lever with your knee so that both hands can be used to guide the fabric.
- **Light** – Is it good enough? Some lights are weak.
- **Low shank, high shank or slant shank feet?** - By far the most popular for home machines is low shank. Adapters are available to go from high shank to low shank. High speed straight stitch machines use high shank feet. The shank type makes a difference if you already have many feet and accessories, in that case you should to get a machine that will let you use your existing feet.
- **Lay down thread spool** – less tension problems then a vertical thread spool, but you can always buy a vertical thread stand.
- **Needle position selection** - This is only for zigzag machines. Lets you control the straight stitch position of the needle, there is usually three Selectable positions; left, right or center. This comes in handy for using accessory feet that require the needle to be in a specific position such as a piecing foot.
- **Numbered thread tension adjustment** – Nice to be able to remember your tension later.
- **Reverse button or reverse lever** – Some types of lever reverse can be locked in even when your hand is not on the lever. This allows you to sew in reverse with both hands on the fabric. Button reverse pops out of reverse as soon as your hand is not on the button. Do you need to sew in reverse with both hands on the fabric? If so you want a lever reverse that stays in reverse.
- **Screw points for guides** – These are small screw holes on the bed of the machine for attaching guides and other accessories.
- **Sound and noise** – Sound OK? Do you like the way the machine sounds? Can you listen to the machine for hours while sewing?
- **Table mount-able** – Some machines are designed for table mounting, others are difficult to table mount. This is only important if you plan on mounting your machine in a table.
- **Top Speed** - What is the top speed in SPM (stitches per minute)? Is the machine fast enough for your use?
- **Twin needle capable** - Some machines can not use twin needles, for example straight stitch machines and zigzag machines with side facing hooks. If you need to do twin needle sewing then

make sure the machine you are getting can use twin needles. If you are going to do a lot of sewing with twin needles you may also want to see if the machine has separate thread paths for each needle, this gives more consistent stitching and tension.

- **Vibration** – Is the machine smooth enough for you?
- **Weight and portability** - If you need to regularly transport the machine this is an issue.
- **Zigzag** – If you need zigzag stitches make sure the machine has them, some older machines are straight stitch only.

Feature checklist - electronic machines

Most of the items on the previous checklist for mechanical machines also apply to electronic machines. The following additional items only apply to electronic machines:

- **Auto back tack** - This may be programmable as to the length and other parameters.
- **Automatic presser foot lifter** - You can program the presser foot to lift automatically after the automatic thread cutter has cut the thread.
- **Automatic thread cutter**
- **Computer interface** - Not many machines have this now, but in the future this will be a common feature. Allows you to program stitch patterns on your computer and have other advance functionality.
- **Decorative stitches** - Does the machine have the stitches you want?
- **Display and controls** - How big and readable is the display? How good and easy to use is the menu system? Is the machine intuitive or difficult to understand? Does it have a built in help system?
- **Low bobbin thread warning** - Tells you when the bobbin thread is low or out.
- **Memory recall of settings** - Can you save and recall your favorite settings?
- **Monogram capability** - The ability to monogram and sew alpha numeric characters. How many fonts are built in to the machine? Can new fonts be added? What is the maximum size of the characters?
- **Needle up/down control** - You can control whether the needle stops in the up or down position.
- **Presser foot awareness** - When you change stitches does the machine tell you which presser foot you should be using?
- **Programmable stitches** – The ability to program your own stitch patterns.
- **Programmable or automatic thread tension** - The thread tension is controlled by the electronics in the machine and the machine can change the thread tension for each stitch type automatically for best results.
- **Start/stop button** - Lets you start and stop the machine using front panel buttons without using the foot controller.
- **Variable speed control** - Lets you control the top speed of the machine and on some models you can run the machine without the foot controller.

Used Machines

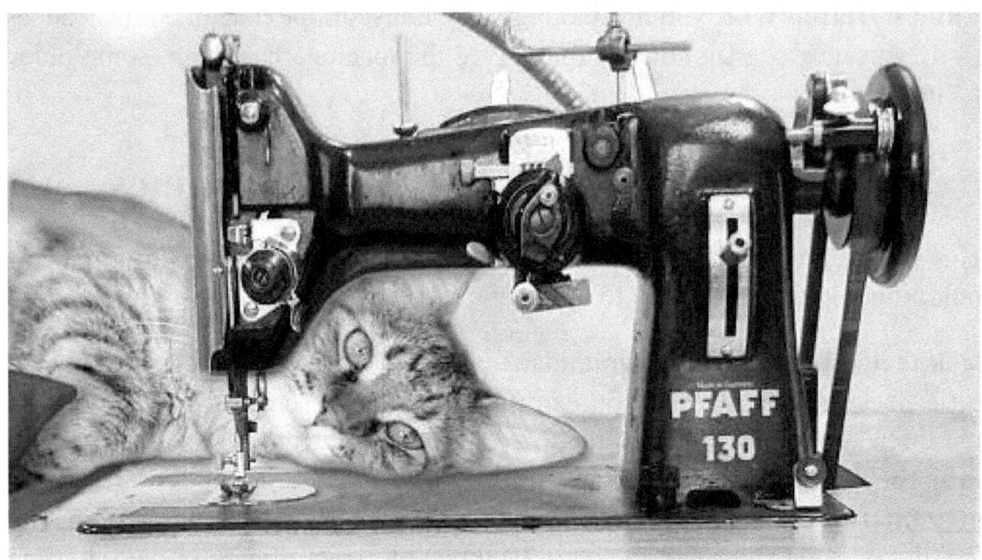

This chapter gives information on a few of the more popular models of used sewing machines that are available. If you are really interested in vintage machines, our companion book "The Sewing Machine Master Guide" covers a large number of vintage machines and also includes detailed information about troubleshooting and repairing vintage machines.

When should you buy a used machine? Used machines are great if you want to save some money on a really heavy duty machine. Keep in mind that you may need to learn how to maintain and adjust the machine or you will have to spend some extra money at the repair shop to have someone else do it, older machines do require maintenance and adjustment!

Criteria

Machines were selected for several reasons as follows:

- Machines that were manufactured in large quantities. These are the machines that you will find most often on the used market.
- Parts are still available on-line through eBay and other sellers.
- Repair shops are familiar with these models. They can be serviced at a reasonable cost.
- Standard bobbins, accessories and feet are used.

Singer 15-91

Description - The 15-91 is the Singer Class 15 machine that is most widely available on the used market today. The 15-91 is a straight stitch only machine. Production of these machines started in the 1930's and ended in 1958. They are available today on the used market for around $100.

Features - Gear drive motor, Adjustable presser foot pressure, dropping feed dogs (using a thumb screw on the underside of the machine), side-loading bobbin, heavy cast iron body and bed (about 29 pounds), all metal mechanical parts, Straight stitch only (no zigzag), needle sizes up to a size 21-(130), heavy duty thread take-up levers and driving mechanisms, needle threads from right to left, Top speed is about 1000 SPM.

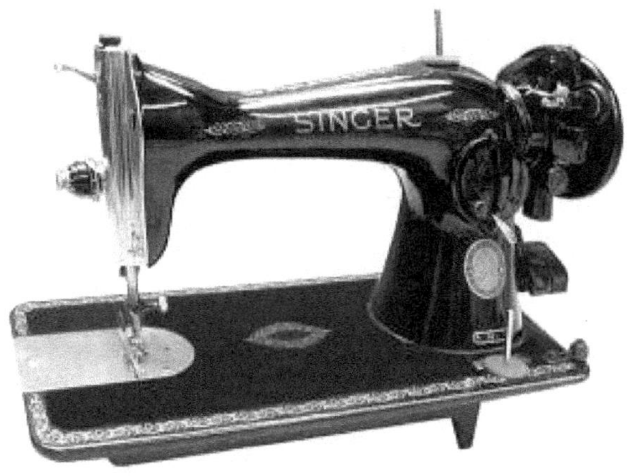

Singer 15-91

Japanese HA-1

Description - Japanese HA-1 straight stitch machines are Japanese copies of the Singer Class 15 models. These machines are of good quality, usually just as good as the Singer Class-15 models, so they are a good choice for a used straight stitch machine and tend to cost less than the Singer machines on the used market. They were sold under many brand names such as: Imperial, Universal, Ideal, Sovereign, Dressmaker, Mercury, National, Brother, Kenmore, Morse, Riccar, Toyota, Free-Westinghouse, etc. There are also similar machines that were produced in other Asian countries (including some recently made in China that can be purchased new), but these non-Japanese made machines are not recommended, the quality is not as good as the 1946-1966 Japanese made machines.

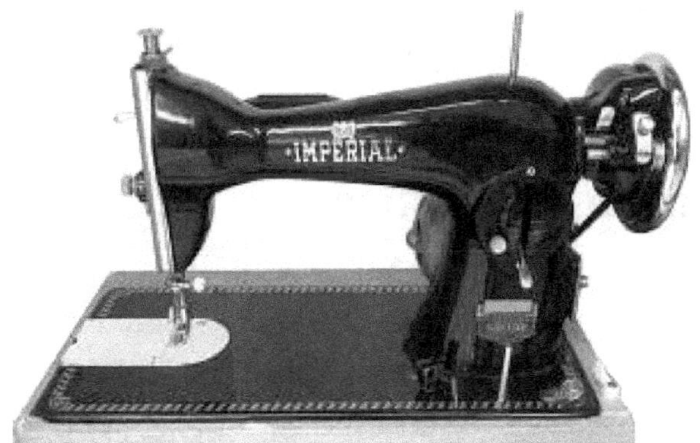

Imperial HA-1

Identification - Most HA-1 machines are not actually labeled as "HA-1", the term HA-1 came into wide spread use after these machines were made and there was no system of model numbers or serial numbers for HA-1 machines. You can identify an HA-1 machine by its appearance and country of manufacture. Most of the HA-1 machines where painted gloss black, but there were also other colors available such as lime green, blue, red, white, etc. HA-1 machines usually have different shaped hand wheels than the Singer models and the motors and bobbin winders look different as well.

Features - Belt drive motor, adjustable presser foot pressure, dropping feed dogs, side-loading bobbin, heavy cast iron body and bed (about 29 pounds), all metal mechanical parts, straight stitch only (no zigzag), needle sizes up to a size 21-(130), heavy duty thread take-up levers and driving mechanisms, needle threads from left to right. Top speed is from 900 to 1300 SPM depending on the machine and motor.

Singer Feather Weight

Popularity - Because of the light weight and "cute" design these machines are very popular with quilters and people that go to sewing classes, but because of this popularity they are expensive. A modern inexpensive or heavy duty machine will do everything that a Feather Weight can do at a much lower cost. That said the fact remains, they are so cute!

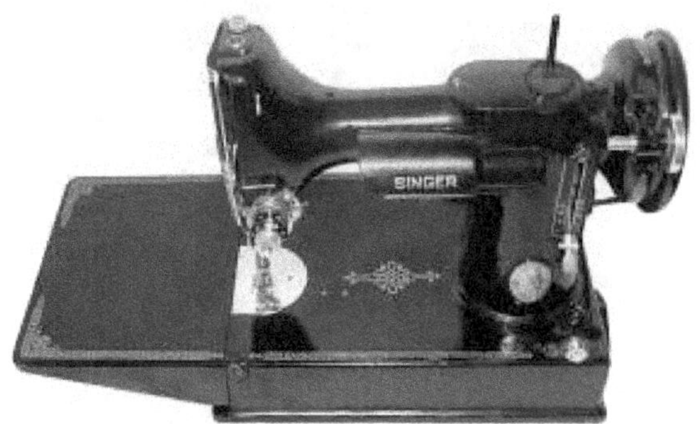

Singer 221

Description - The Singer Feather Weight (Singer model 221) is a small sized machines (about 25% shorter than a full sized machine) with light weight aluminum body. It weighs only 11 pounds. Usually comes with a cute carrying case that looks like a box with a handle. The model 221 was produced from 1933-1970, over 2 million were made. The left side of the bed is hinged and tilts up to make the machine smaller for storage and transport. Cost on the used market varies from about $100 to over $1000 depending on condition and your luck.

Features - The 221 features a belt drive motor, adjustable presser foot pressure, side-loading bobbin, straight stitch only (no zigzag). Needle sizes up to a size 18-(110), needle threads from right to left. Top speed 700 to 1000 depending on the motor.

Japanese cast iron zigzag machines

Description - The Japanese cast iron zigzag machines are real workhorses, they are essentially zigzag variations of the Japanese HA-1 straight stitch machines with a more modern appearance. Many millions of these machines were made so they are widely available. Some people call them "HA-1 zigzag machines". These machines are of good quality and are relatively easy to maintain so they are a good choice for a used heavy duty zigzag machine. They were sold under many brand names such as White, Necchi-Alco, Morse, Alco, Visetti, Dressmaker, Montgomery ward, Brother, Kenmore, Riccar, Toyota, American Beauty, etc.

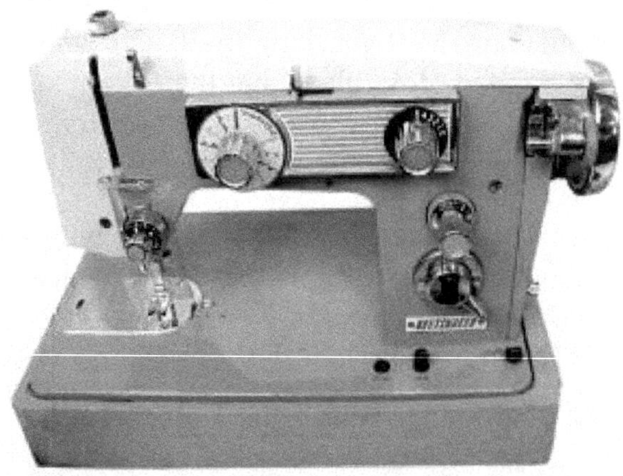

Dressmaker HA-1 - gray

Cost and reliability - Used prices for these machines are from $5 through $125, but more important than getting the lowest price possible is to get a machine in good condition because you can waste more time than the money saved trying to fix a machine with major issues. That said, these machines are tough, it is very rare to see one that can not be serviced and brought back to great working condition unless it has been dropped, damaged or submerged in water. Usually they just need some oiling, TLC and adjustment to sew perfectly.

Identification and buying tips - Identification of these machines is easy, if it is heavy (cast iron, over 32 pounds) and it was made in japan then it is one of these machines! The brand name means very little, it is very hard to determine which manufacturer made any given machine, furthermore all of the manufacturers that made these machines were good so it really does not matter much who made the machine. More important than the brand or manufacturer is the condition and features of the specific machine you are looking at. If possible test sew the machine.

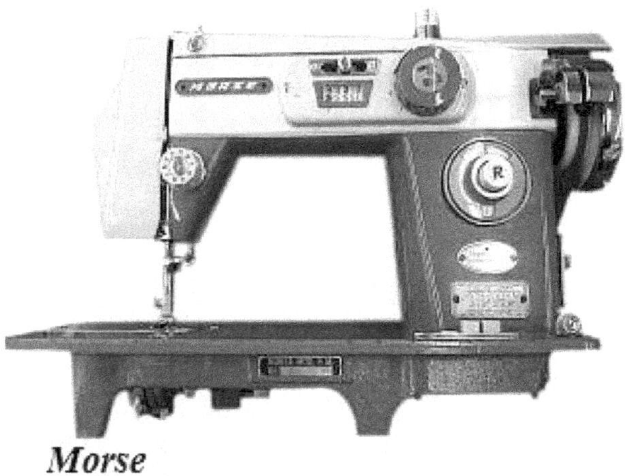

Morse

Colors and styling - Japanese cast iron zigzag machines are available in a wide array of colors and styles ranging from ultra conservative to wild retro and art deco designs that were inspired by stream lined locomotives and 1950's automobiles.

Features - Belt drive motors, adjustable presser foot pressure, dropping feed dogs, heavy cast iron body and bed (over 32 pounds), all metal mechanical parts, zigzag, needle sizes up to a size 18-(110), heavy duty thread take-up levers and driving mechanisms. Top speed is from 900 to 1300 SPM depending on the machine and motor.

How To Use A Sewing Machine

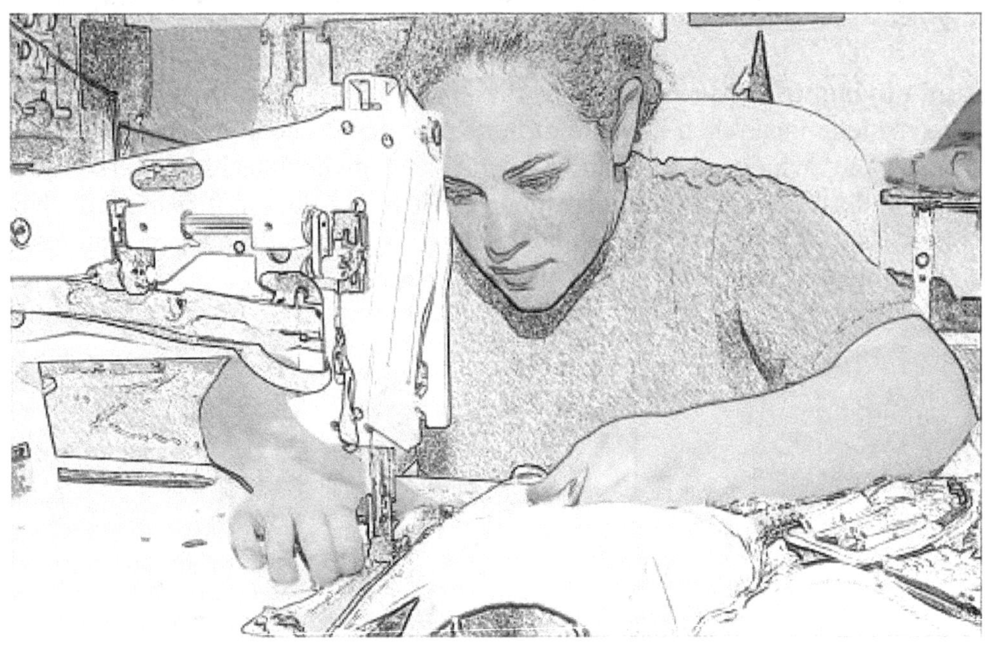

Before you start sewing

For general purpose sewing use a size 14 needle. Most new home machines come with a size 14 needle pre-installed in the machine. Use a polyester general purpose sewing thread like one of the following;

- Coats & Clark - Dual Duty XP
- Coats & Clark - All Purpose
- Gutermann - Sew-All
- American & Efird - Excell.

Tip - Reading glasses can be used as a handy magnifier for threading the machine and inspecting stitch quality (even if you don't need them for reading). Reading glasses are better than a magnifying glass because they allow the use of both of your hands and give you better depth perception because each of your eyes has its own lens. Reading glasses are now selling in 3 packs at Walmart and on Amazon for $7.99

Winding the bobbin

Before we thread the machine we must wind thread onto the bobbin. If you have the owners manual for your machine you should follow the instructions for your specific model of machine. In general bobbin winding is done as follows:

1. Place an empty bobbin onto the shaft of the bobbin winder. Never wind thread onto a bobbin that already has some thread on it, this can cause tension problems and thread knots. If there is some thread on the bobbin then remove it first.

2. For bobbin winding the bobbin winding clutch must be disengaged. If the clutch is properly disengaged the needle of the machine will not go up and down when the handwheel is turned. The motor can then be used to wind the bobbin without the whole machine running and vibrating.

On newer machines the bobbin winding clutch is disengaged by simply moving the bobbin towards the bobbin winding stop.

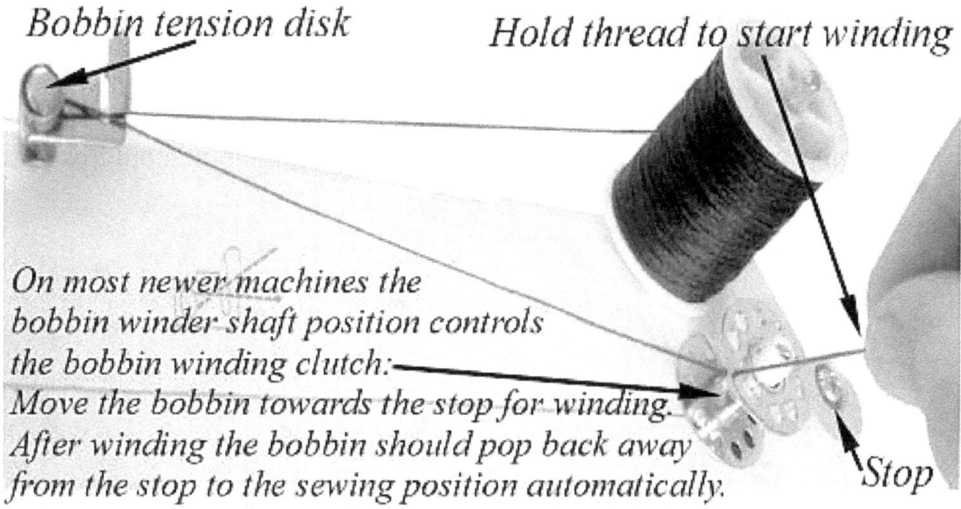

If your machine has the bobbin winding clutch in the center of the hand wheel, you will need to disengage the clutch before winding the bobbin. This is done by turning the knob counterclockwise with one hand while preventing the handwheel from moving with your other hand. After bobbin winding make sure to turn the knob clockwise to re-engage the clutch for sewing. On this type of machine you will also have to engage the bobbin winding release latch, this latch will engage the bobbin winder so the bobbin will turn. On some machines this latch is located next to the bobbin and looks like a lever that moves towards the center of the bobbin. On other machines the bobbin shaft moves to control this latch.

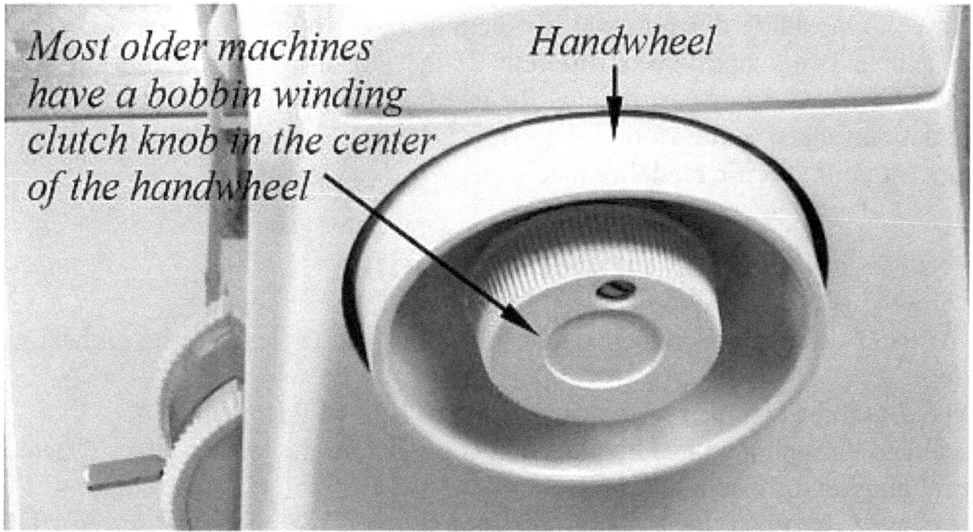

3. Place the thread spool on the appropriate spool pin. This may be the same spool pin that is used for sewing, but some machines have another separate spool pin that is used only for bobbin winding.

4. Pass the thread from the thread spool through any thread guides that may be on the way to the bobbin winding tension disk. The bobbin winding tension disk is not the same as the main tension disk used for sewing, it is generally much smaller and may look like a small button on a bracket. After passing the thread through the bobbin winding tension disk, pass the thread through any remaining thread guides and then to the bobbin.

5. Pass the thread through one of the holes or slots in the side of the bobbin and hold the thread as you start to wind the bobbin (see the first picture in this section). You can let go of the thread after the bobbin winds for a few turns.

6. Press the foot pedal of the machine to start the bobbin turning.

7. The bobbin should wind evenly. If it does not wind evenly you should figure out why. Some people guide the thread during bobbin winding with their fingers to make sure that the thread winds evenly.

8. Do not wind too much thread onto the bobbin. This will cause tension problems or thread breakage because the thread extending beyond the sides of the bobbin will contact the bobbin case and stop the bobbin from freely turning in the bobbin case. A correctly wound bobbin should have the thread level below the sides of the bobbin. See the picture below.

Overfilled *Properly filled*

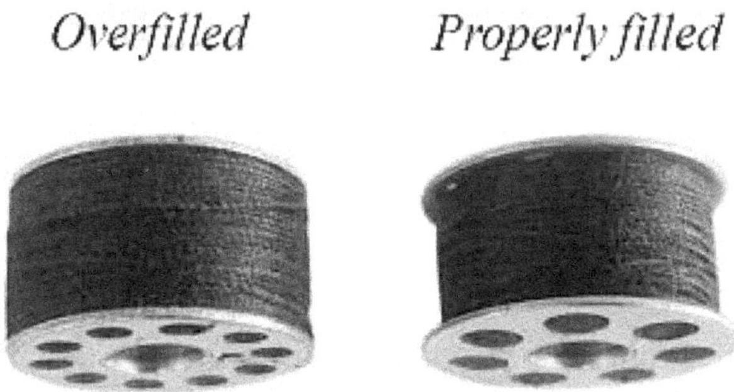

9. Don't forget to re-engage the bobbin winding clutch after you finish winding your bobbins.

Threading diagrams

It is best to follow the threading diagram and instructions in the owners manual for your specific make and model of machine. Every machine is different. If the threading instructions for your machine are confusing or you do not have an owners manual the instructions below may help. The instructions below are general instructions that should work for most machines. You can also search the Internet for a threading diagram for your specific model of machine.

Upper threading

1. Rotate the hand wheel counterclockwise to raise the take up lever to its highest point. Raise the presser foot.
2. Set the spool of thread on the spool pin at the top or back of the machine.
3. Pass the thread through the thread guides that come before the tension disks. There may be one or more depending on the machine.
4. Pass the thread between the tension disks and over the take up spring. There are many configurations of tension disks and take up springs, you may need to do a search on the Internet for a threading diagram for your specific model of machine if you are having trouble with this step (see close-up pictures below).
5. Pass the thread through the take up lever.
6. Pass the thread through the thread guides on the way to the needle, there may be one, two or more of them.
7. Thread the needle. Most machines thread from front to back or from left to right, but some older Singers like the 15-91 thread from right to left. You may need to do a search on the Internet for a threading diagram for your specific machine if you are not sure.
8. Pull out about 5 inches of thread from the needle and pass the thread under the presser foot and to the back of the machine.

Many newer home machines have a slotted thread path. Machines that have a slotted thread path have the tension disks mounted inside the slot where they can not be seen. This picture shows the threading for a generic home machine with a slot type threading:

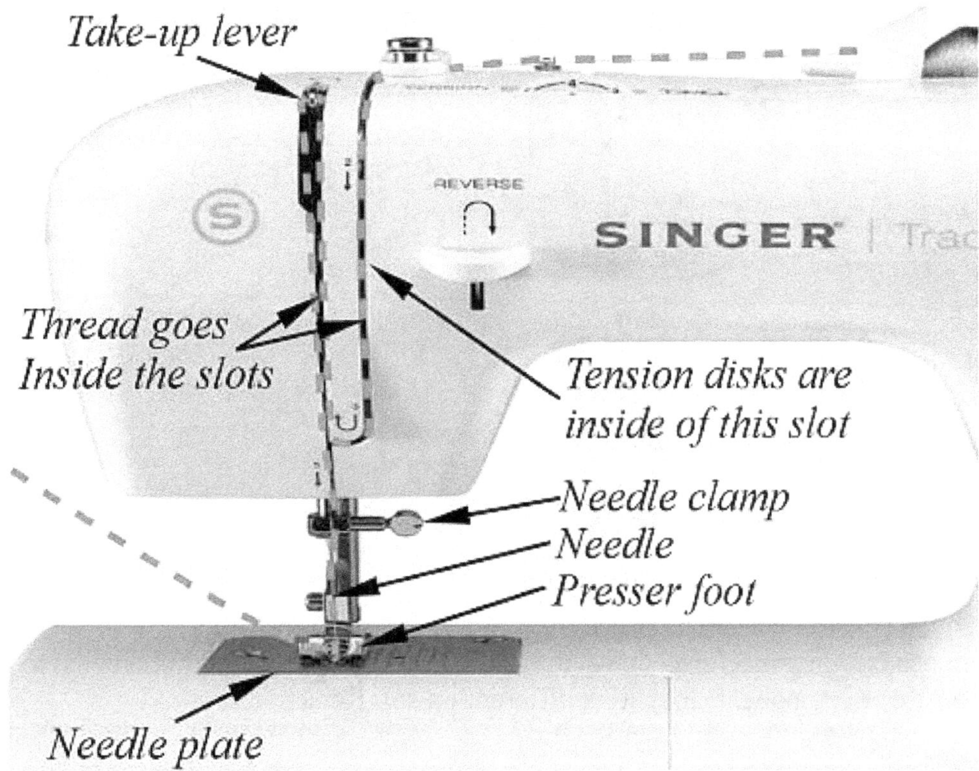

Most older machines and heavy duty machines have a conventional thread path. The next picture shows the threading for a conventional thread path:

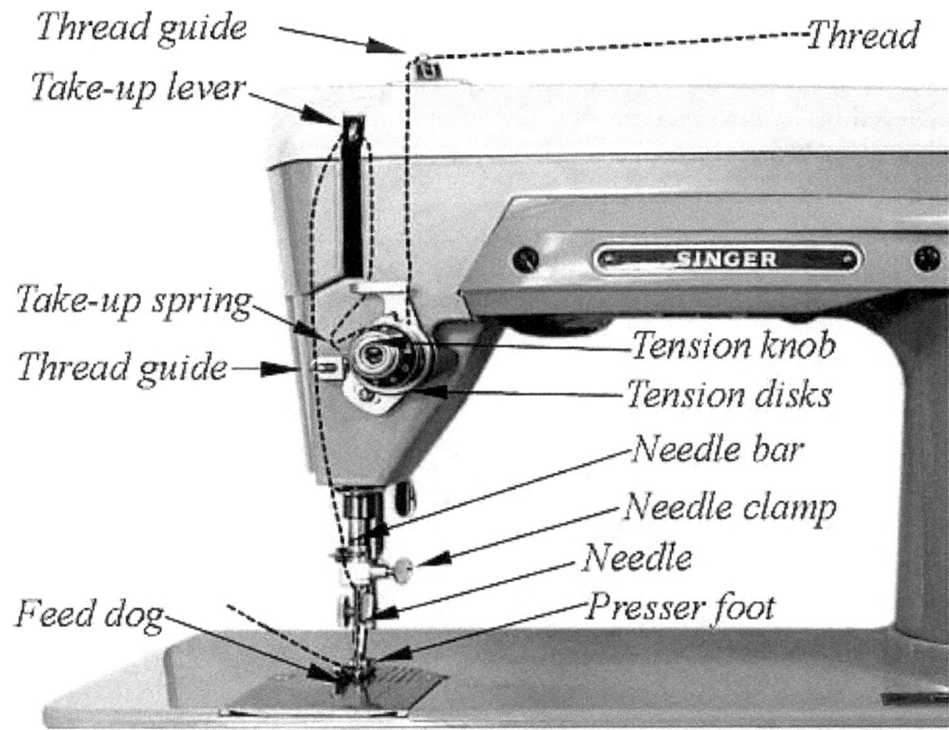

Below is a close-up of the tension knob threading. It is important to note that the thread goes around and under the tension knob after entering the tension disks and then exits the tension disks at the tension disk exit thread guide.

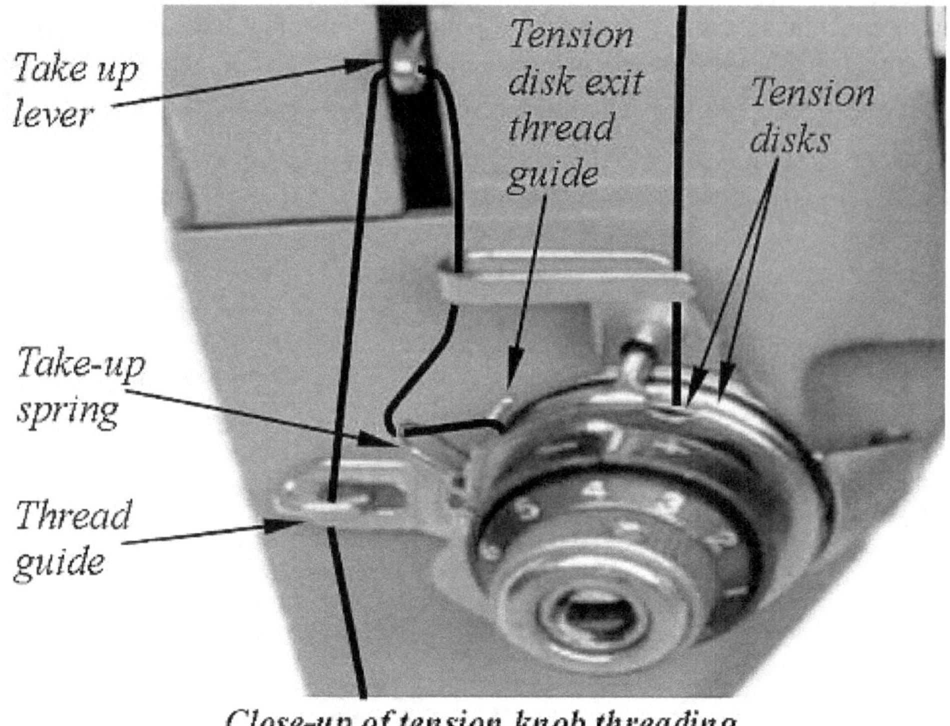

Close-up of tension knob threading

All general purpose sewing machines have a similar thread path and components to the machines pictured, however these components may look different on some machines.

Lower threading

If you have not already done so during the upper threading then rotate the hand wheel counterclockwise to raise the take up lever to its highest point. Raise the presser foot.

Machines with top loading bobbins - If your machine has a top loading bobbin, drop the bobbin into the bobbin case as pictured in the next picture. Pull the thread through the slot and under the bobbin tension spring. Pull out about 5 inches of thread.

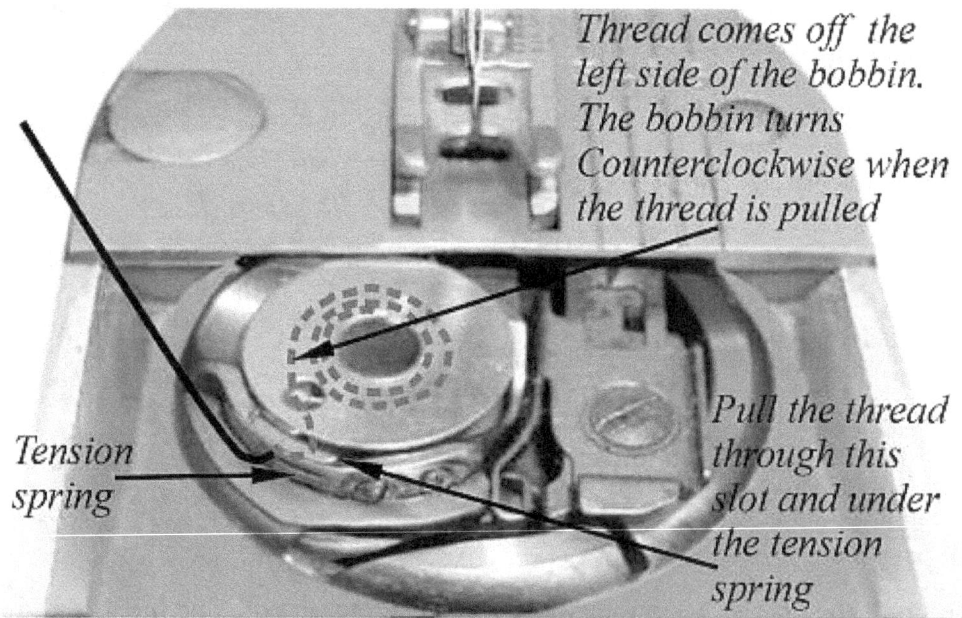

Bobbin threading for a top loading machine

Machines with vertical bobbins - If your machine has a vertical bobbin and a removable (HA-1 type) bobbin case follow these steps;

1. Remove the bobbin case from the machine by pulling the latch on the front of the bobbin case.
2. Insert the bobbin into the bobbin case as pictured in the diagram below.
3. Pull the thread through the slot and under the bobbin tension spring. Make sure that the thread is all the way under the tension spring as pictured.
4. Pull out about 5 inches of thread. You should feel an even and consistent tension on the thread.
5. Insert the bobbin-case into the machine. It should snap in place when it is inserted properly. If you use the latch to grab the bobbin-case while inserting it into the machine the latch mechanism will hold the bobbin from falling out of the bobbin-case.

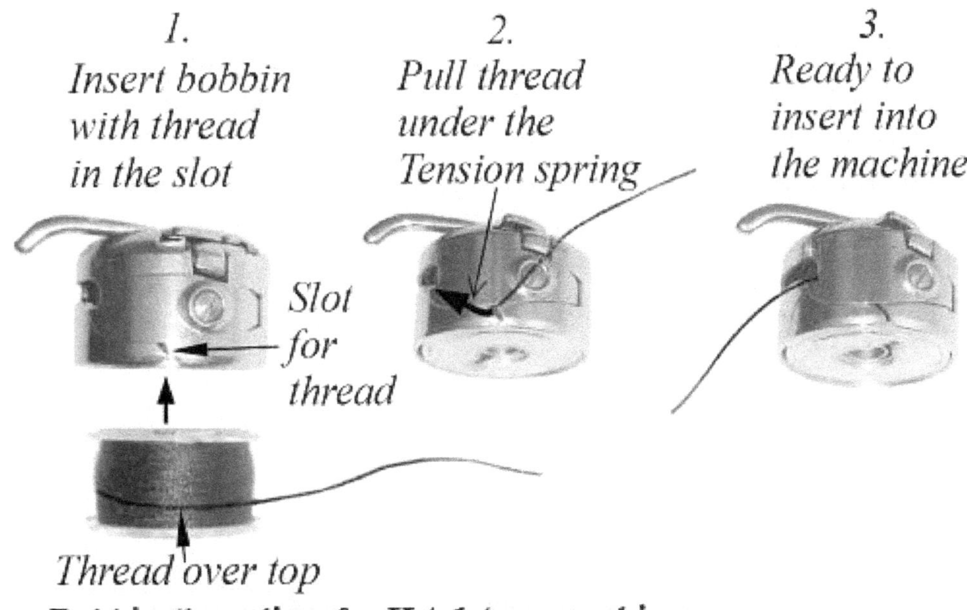

Bobbin threading for HA-1 type machine

Below is a picture of the HA-1 bobbin case after it has been inserted into the machine.

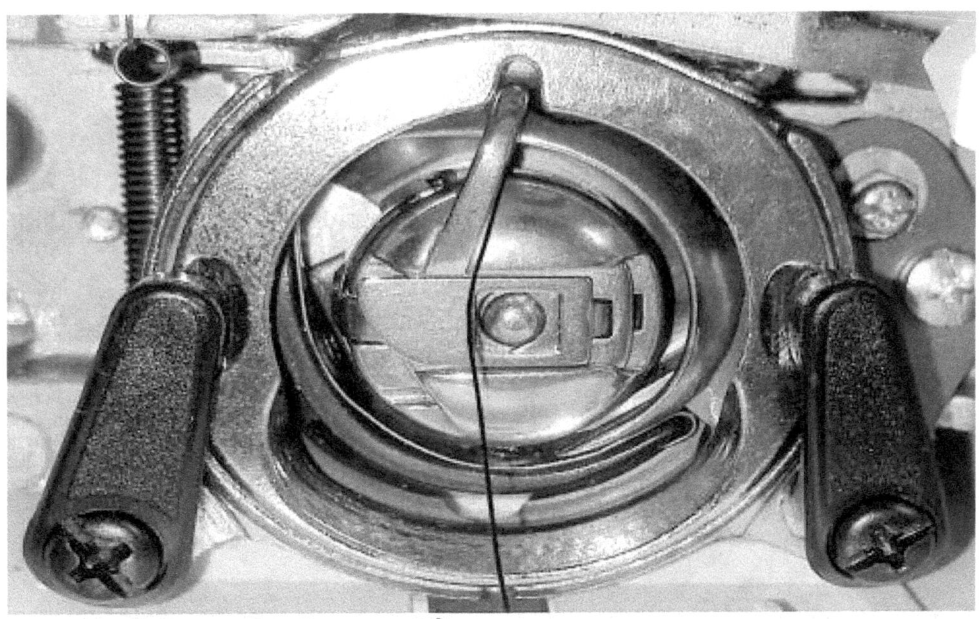

Bobbin-case in the machine

Raising the bobbin thread

On most mechanical machines you must raise the bobbin thread when you first thread the machine and subsequently when you replace the bobbin after it runs out of thread. Raising the bobbin thread means to bring the bobbin thread up through the needle hole. For electronic machines you should follow the instructions in your owners manual, some electronic machines raise the bobbin thread automatically. For most mechanical machines raising the bobbin thread is done as follows:

1. The presser foot needs to be moved to the down position (lower the presser foot).

2. Start with the needle in the up position, hold the needle thread with your left hand and pull gently towards the front of the machine. Do not pull the thread hard enough to bend the needle or to pull out any thread.

3. Turn the hand wheel toward you (counter clock wise) one full turn with your right hand while you continue to hold and lightly pull the needle thread with your left hand. The needle should start and stop in the up position (this will be one full turn of the hand wheel).

4. As you are doing step 3 you will notice that as the needle thread comes out of the hole in the needle plate (as the needle raises) it will be looped with the bobbin thread. The needle thread will have pulled the bobbin thread up through the hole in the needle plate.

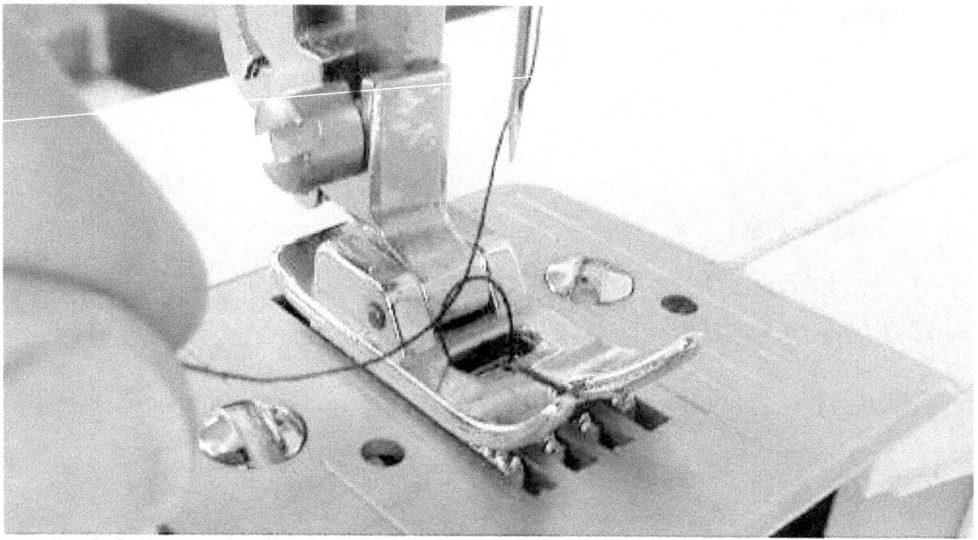

Raising the bobbin thread

5. Next pull out the loose end of the bobbin thread. Now there should be about 4 or 5 inches of thread coming out from both the bobbin and needle. Unloop the bobbin thread from the needle thread if the two are entangled.

6. Raise the presser foot and place both threads under the presser foot with the threads drawn towards the back of the machine and you are ready to start sewing.

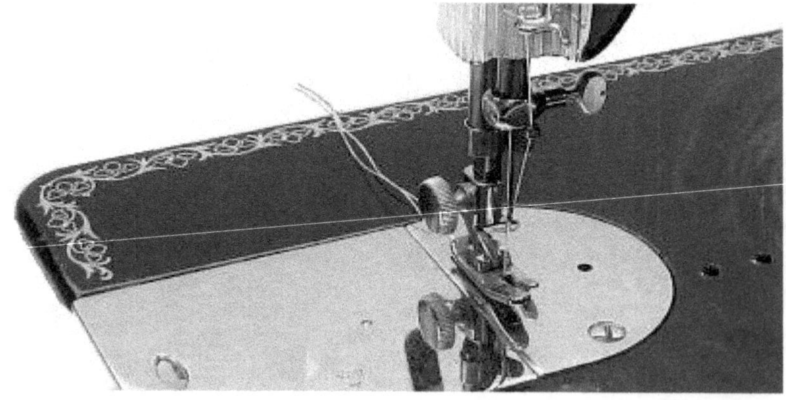

Setting the thread tension

All sewing machines have recommended tension settings that are given in the owners manual. To start, use the recommended tension settings for your machine. Some machines have a mark or special color on the tension dial to indicate a default tension for the machine. A default tension is used as a starting tension and should produce a usable stitch in a medium weight fabric when using general purpose polyester thread and a medium stitch length.

If you do not have an owners manual for your machine and your machine does not have default markings on the dial, start with a tension setting that is 1/3rd of the maximum. On most machines the tension dial goes from 0 to 9 and this setting will be a 3 or 3-1/2.

Setting the stitch type and stitch length

For most fabrics - use a straight stitch with a medium stitch length of about 2 mm (or 12 SPI if you have an older machine that uses SPI instead of millimeters). You can use this stitch for most of your sewing. SPI means Stitches Per Inch.

For stretch fabrics - If you have a zigzag machine then use a medium width zigzag stitch with a medium stitch length. If you have a straight stitch machine then use a short stitch length of 1mm (26 SPI) and loosen your needle thread tension somewhat to allow the seam to stretch.

Sewing with older mechanical machines

1. Position the fabric under the presser foot at the spot where you want to start your stitch and lower the presser foot.
2. Using the fingers from your left hand grab the thread tails and pull them toward the back of the machine, but don't pull hard enough to pull more thread out or bend the needle, just pull hard enough to prevent the thread tails from being pulled down into the machine at the start of sewing. If you allow the thread tails to be pulled into the needle hole of the machine at the start of sewing, most older mechanical machines will make an ugly knot of thread on the underside of the fabric at the start of the seam. Once the machine starts and the first few stitches have been made you can let go of the thread tails and sew with no worries. Holding the thread tails at the start of the stitch becomes second nature after a while and you won't even think about it while you are doing it.
3. Lightly place the fingers of your right hand on the fabric in front of the presser foot (about 2 inches or more in front of the presser foot) to guide the fabric through the machine.
4. Using your foot, press on the foot controller to start the machine. You are now sewing!
5. After the machine has made the first few stitches you can let go of the thread tails you are holding in your left hand and use your left hand to help guide the fabric through the machine.
6. Do not push or pull the fabric through the machine. Let the machine move the fabric. Only use enough effort to guide the fabric in the direction you want to sew. The machines feed dogs will only move the fabric when the needle is in the up position. This is important because if you try to push or pull the fabric through the machine you will move the fabric when the needle is in the down position and you will bend the needle and possibly cause the needle to hit the needle plate and break. If you are sewing a thick seam you may need to help the machine move the fabric, but in this case you should advance the machine yourself one stitch at a time using the hand wheel. When the needle is up (and only when the needle is up) push the fabric ahead to the next stitch position. Then rotate the hand wheel one turn to make another stitch. Repeat this to make additional stitches until you are over the thick seam or other hard to sew areas. Remember never try to move the fabric yourself when the needle is in the fabric or the machine is running.
7. When you are done sewing then use the hand wheel to rotate the needle past the highest point and

then slightly down before raising the presser foot to remove the fabric from the machine. This is done so that the thread will release from the hook mechanism of machine.

8. When you remove the fabric from the machine pull it out towards the back of the machine and cut the thread that is still attached to the fabric leaving about four inches of thread coming from the needle and bobbin so that the machine is ready for additional sewing.

Sewing with newer electronic machines

Keep in mind that every manufacturer of electronic machines does things differently so it is important to also learn how these steps are done from the owners manual for the specific model of machine that you are using. If the instructions in the owners manual for your machine are different from the steps below then follow the instructions in your owners manual.

1. Raise the needle to the highest position if it is not already up. You can raise the needle by pressing the needle-up button or by rotating the handwheel.
2. Position the fabric under the presser foot at the spot where you want to start your stitch and lower the presser foot.
3. Lightly place the fingers of your right hand on the fabric in front of the presser foot (about 2 inches or more in front of the presser foot) to guide the fabric through the machine.
4. Using your foot, press on the foot controller to start the machine. You are now sewing!
5. Do not push or pull the fabric through the machine. Let the machine move the fabric. Only use enough effort to guide the fabric in the direction you want to sew. The machines feed dogs will only move the fabric when the needle is in the up position. This is important because if you try to push or pull the fabric through the machine you will move the fabric when the needle is in the down position and you will bend the needle and possibly cause the needle to hit the needle plate and break. If you are sewing a thick seam you may need to help the machine move the fabric, but in this case you should advance the machine yourself one stitch at a time using the hand wheel. When the needle is up (and only when the needle is up) push the fabric ahead to the next stitch position. Then rotate the hand wheel one turn to make another stitch. Repeat this to make additional stitches until you are over the thick seam or other hard to sew areas. Remember never try to move the fabric yourself when the needle is in the fabric or the machine is running.
6. When you finish the seam and stop the machine the needle will be up or down (depending how your machine is programmed). If the needle is in the down position then you will need to press the needle-up button to cause the needle to move to the up position so that you can remove the fabric from the machine.
7. If your machine has an automatic thread cutter then you may need to press the thread cutting button to cut the thread at the end of the seam.
8. Raise the presser foot and remove the fabric from the machine.
9. If you do not have an automatic thread cutter then manually cut the thread that is still attached to the fabric so that there is about four inches of thread coming out of the needle and bobbin.

Checking the tension

Use a piece of scrap fabric to sew a test seam. Inspect the stitches on the top and bottom of the fabric for balanced tension. It is best to use opposite colors of thread and fabric for test sewing so that the stitches are clearly visible. You may want to use different colors for the needle thread and bobbin thread like in the diagram below.

In the diagram below you are looking at a seam from the side as if the fabric was cut along the seam line. This is called a cross-sectional view. The fabric appears much thicker than real fabric is, this is done so that you can clearly see how the needle thread crosses the bobbin thread at the lock point. The tension is balanced when the lock point is close to the center of the fabric.

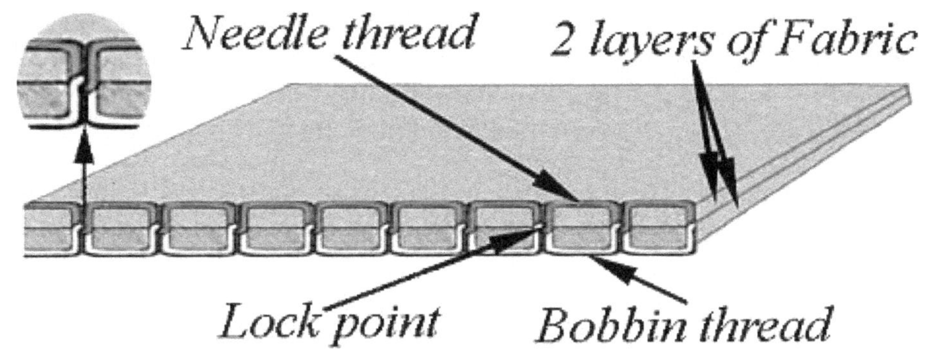

Check your test seam to see if the lock point is in the center of the fabric like in the above diagram. Most machines do not have perfect tension, so it is OK if the lock point is not exactly in the center of the fabric as long as it is not off as badly as shown in the next diagram.

This diagram shows what happens if the needle thread tension is too tight, correct or too loose.

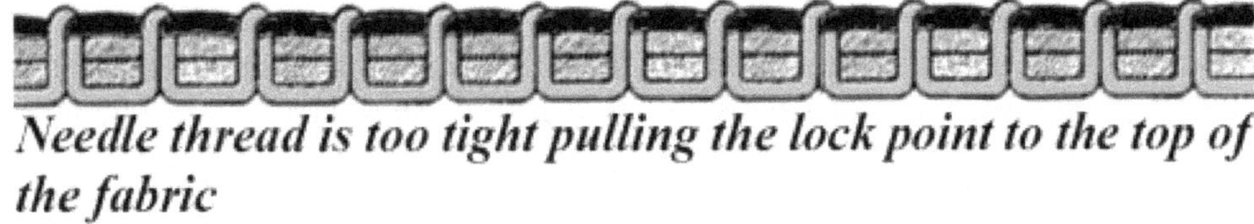

Needle thread is too tight pulling the lock point to the top of the fabric

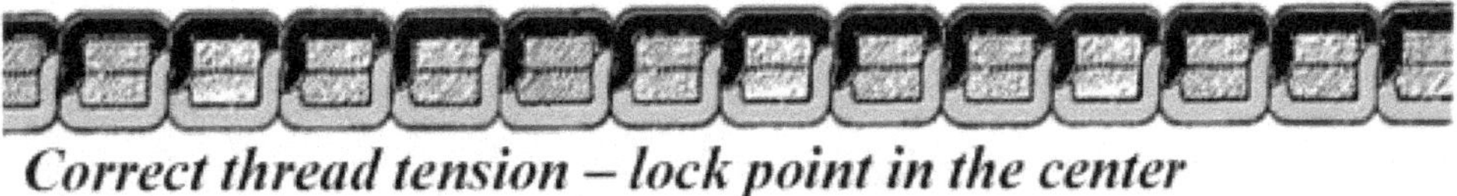

Correct thread tension – lock point in the center

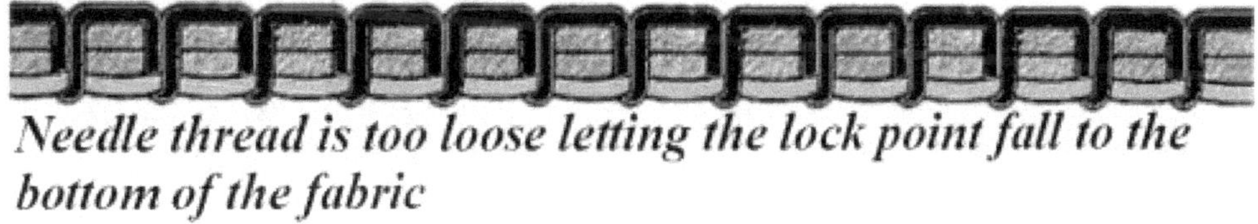

Needle thread is too loose letting the lock point fall to the bottom of the fabric

If your tension is not correct then adjust the needle thread tension of your machine a small amount and test sew again. If your machine has a tension knob that goes from 0 to 9 then try adjusting your tension by 1 step or 1/2 step at a time. If your tension is still not correct or you want to read more about adjusting tension then go to the chapter "Adjusting tension".

Tips and Ideas

Reinforcement - You may want to reinforce the ends of your seams. To do this sew about ½ inch (12mm) in the opposite direction (put the machine in reverse) at the start and end of your seams. Some sewing machines can do this automatically, they have a special button or can be programmed to do this at the start and end of every seam. In this case it is called automatic back tacking or automatic tacking.

Holding the thread tails at the start of sewing - Not all machines require the thread tales to be held at the start of sewing, but most older machines or machines with an HA-1 style bobbin-case do. Here is how it is done; At the start of sewing use the fingers from your left hand to grab the thread tails and pull them toward the back of the machine, but don't pull hard enough to pull more thread out or bend the needle, just pull hard enough to prevent the thread tails from being pulled down into the machine at the start of sewing. If you allow the thread tails to be pulled down the needle hole of the machine at the start of sewing, most older mechanical machines will make an ugly knot of thread on the underside of the fabric at the start of the seam. Once the machine starts and the first few stitches have been made you can let go of the thread tails and sew with no worries. Holding the thread tails at the start of the stitch becomes second nature after a while and you won't even think about it while you are doing it.

Don't push or pull the fabric - Let the feed dogs do their work, their job is to move the fabric through the machine. Your job is to guide the fabric so that the stitching is in the correct place.

Need maximum seam strength? - If you need maximum strength you may not need to use thick thread if your machine can do a double or triple stitch. In this stitch the machine reverses on every stitch and does it two or three times before moving on the the next stitch thereby making the seam much stronger while still using a regular thread size.

Head start for heavy fabric or thick seams - If you are sewing heavy fabric or thick seams then give your machine a "head start" by rotating the hand wheel towards you (You always turn the hand wheel counter-clockwise and never clockwise on modern sewing machines) to lower the needle down until it is just sicking through the fabric (and then stop rotating). This is done to give the machine a head start, then when you start the machine it has time to build up speed before the needle starts its next stitch. Without doing this "head start" the machine may not have enough power to drive the needle through the thick fabric and will stall and make a buzzing sound. After a while you will know when this "head start" is needed.

Very thick seams - If you are sewing through a very thick seam (denim jeans, jackets, belts, cases or similar) then use manual power (rotate the hand wheel of the machine yourself) instead of motor power to go through the seam as follows:

- Rotate the hand wheel (counter-clockwise) so that the needle goes through the fabric and continue to rotate the hand wheel until the needle is in the up position (out of the fabric).
- Next move the fabric to the next stitch position by hand if the machine was not able to properly move the fabric to the next stitch position by itself. Only do this with the needle in the up position (out of the fabric) or you will bend the needle. You may need to rock or rotate the fabric slightly from side to side to get it to move, after you do this a few times you will get proficient at it. You may need to raise the presser foot release lever slightly to lessen the force on the presser foot while you are moving the fabric.
- Repeat step 1 and 2 again and again until you are past the thick seam and can sew normally again.

Corners - Stop the machine at the location where you want to make the corner. Rotate the hand wheel so that the needle is down at its lowest position going through the fabric (you can set electronic machines to do this automatically). Raise the presser foot and then rotate the fabric using the needle as a pivot point so that you will now be sewing in the new direction you want to go. Do this gently so as not to bend the needle or scrunch up the fabric. Lower the presser foot and start sewing again.

Curves - To sew a tight curve you can stop the machine and slightly pivot several times or use the hand wheel to run the machine slowly to achieve better control and a smooth curve instead of trying to sew it at one time. Remember before pivoting on the needle make sure that the needle is all the way down and try to put as little pulling force on the needle as you can.

Stretchy fabric - To sew stretchy fabric like knits or spandex you may need to tension the fabric while sewing. This is done by pulling equally at the front and back of the fabric while sewing to stretch the fabric that is going through the machine. However, even though you are stretching or "applying tension" to the fabric you must let the feed dogs move the fabric through the machine. Be careful to keep the tension equal and do not pull or push the fabric through the machine. Alternatively you can use stabilizer (see next item).

Stabilizer - Use stabilizer to help with even feeding of stretch fabrics or very thin fabrics. Stabilizer is a paper like material that you use over the fabric (or under or on both sides). The stabilizer stops the fabric from stretching or bunching. Stabilizer paper can be purchased from fabric stores or you can use some other thin paper like tissue paper, news paper or phone book paper. The stabilizer paper that you buy works better because it is easier to remove after the seam is made and leaves fewer bits of paper stuck in the seam that have to be picked out with tweezers after sewing. Actually if you can use a shorter stitch length then sometimes the closely spaced needle holes will cut the paper stabilizer allowing clean removal.

Don't sew over pins - Sewing over pins can bend or break your needle and can damage your machine. If you do pin your fabric make sure the pins are well away from the edge of the fabric or remove the pins as you sew and before they get to the needle area of the machine.

Thick seam equalizer - To sew over thick seams you can use a non-pivoting presser foot or a locking pivot presser foot to help the fabric feed correctly. See the chapter "Presser Feet" for more info on locking presser feet. Alternatively you can use a spacer (like a Jean-a-ma-jig) under the presser foot to help the machine feed the fabric through the seam.

Sewing slow - Most sewing machines don't go slow very well. If you need to go slower than your machine likes to go under its own power then you should use manual power (rotating the hand wheel of the machine with hand power). Do this whenever you need the machine to sew very slowly with a great deal of control. Using this technique you can sew very accurately in problem areas and control the location of each stitch precisely. This is handy for sewing around corners, installing zippers, lapels, belt loops, sewing very small details, etc. Always rotate the hand wheel toward you only (counter clock wise) and never the other direction. See the section "Thick Seams" above for another description of manually advancing the machine. On some electronic machines it is possible to advance the machine automatically one stitch at a time by pressing a button.

Skill - It's mostly skill that gets good results, not the machine! The machine is just a tool, you control what the machine does.

Mash on it - You don't have to start slowly or stop slowly unless you are sewing thick materials or multiple layers. If you are sewing light to medium fabrics and you are sewing long seams and need to go fast then mash on it! Sewing machines are designed for full power start-and-stop sewing.

Adjusting Tension

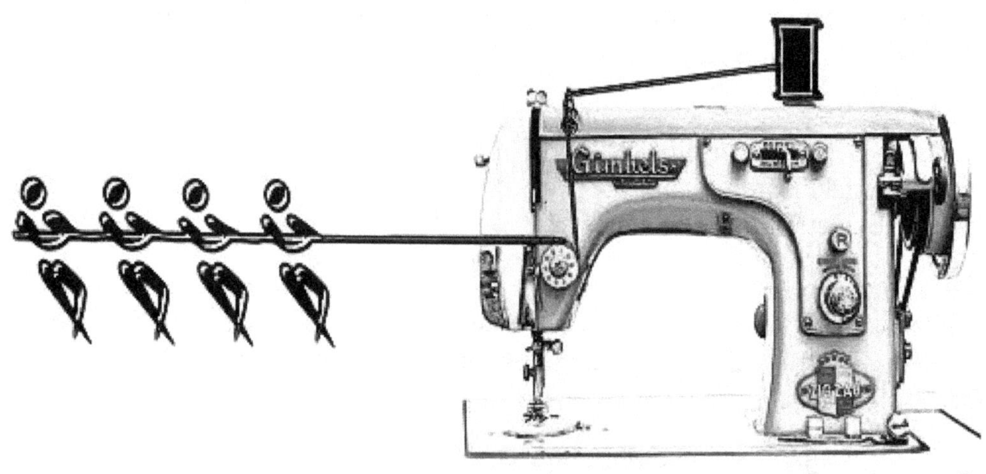

About tension

For a sewing machine to function properly the thread must be under the correct amount of tension. To provide the proper amount of tension a sewing machine has adjustable disks in the thread path. These disks pinch the thread to create tension. You can demonstrate this by squeezing thread between your two fingers. As you squeeze your fingers you can control the tension on the thread depending on how hard you squeeze.

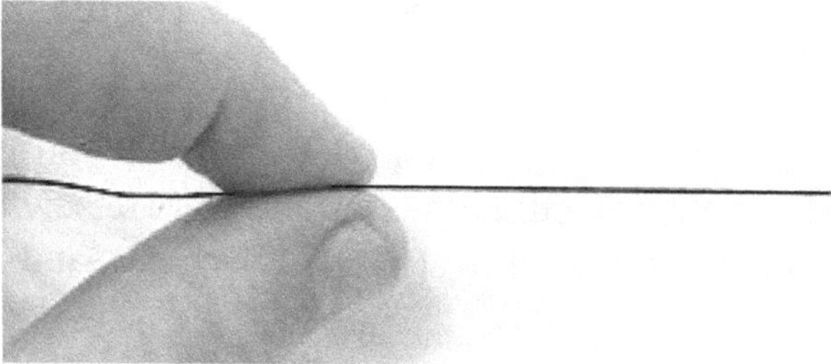

The tension disks of a sewing machine work the same way except that the tension is adjusted by a knob or screw instead of by applying finger pressure. Below is a picture of some older style tension disks. New machines have the tension disks hidden inside a slot that the thread goes through on the front of the machine.

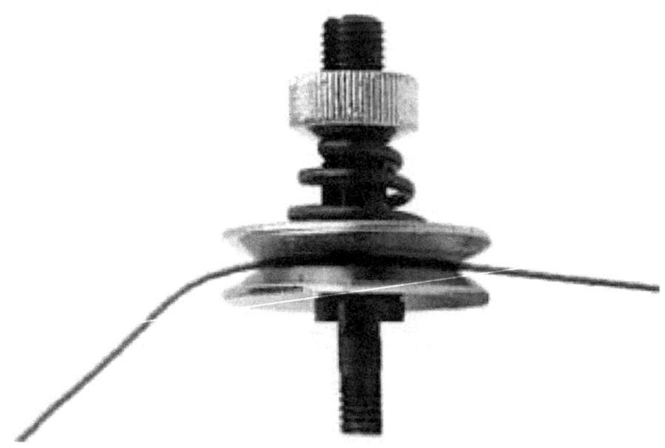

Tension adjustment knobs and screws

There are two tension adjustments on a sewing machine, one for the needle thread and one for the bobbin thread.

The needle thread tension adjustment is controlled by a knob or dial on the upper part of the machine. Most tension adjustments will be made by adjusting the needle thread tension and not the bobbin thread tension. The bobbin thread tension can be left alone most of the time. The next picture shows a needle thread tension dial with numbers from 0 to 9 with 0 being almost no tension and 9 being the greatest amount of tension possible.

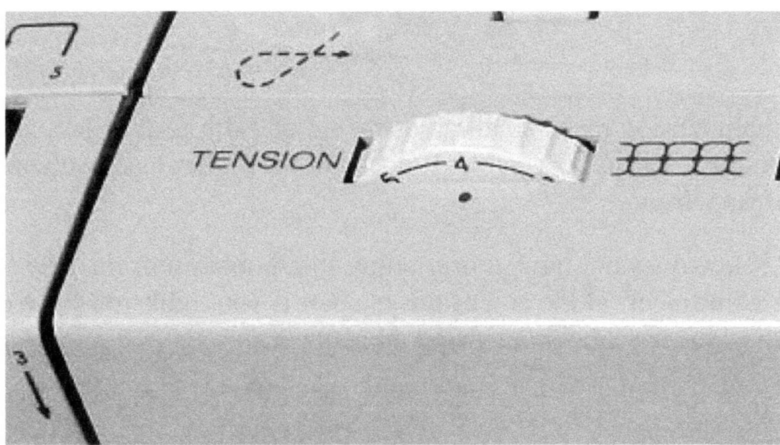

An example of tension knob is shown in the picture below.

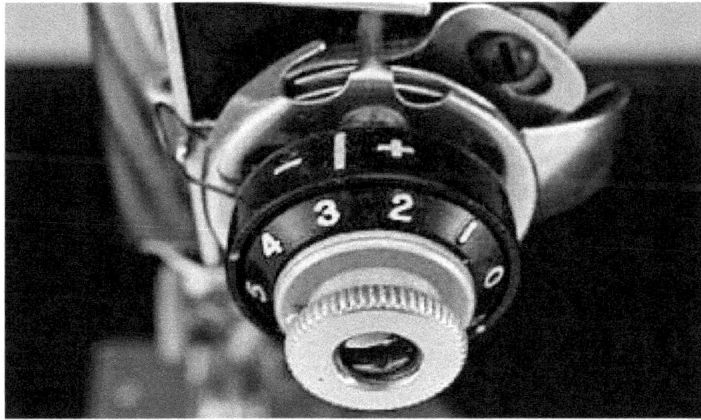

The bobbin thread tension adjustment is controlled by a small screw located on the bobbin case or on some newer machines it is not adjustable and is preset at the factory. Not all bobbin cases look like the one in the picture below, the bobbin case on your machine may look much different but the function is the same no matter what they may look like.

On machines that have adjustable bobbin tension, to adjust the bobbin tension turn the screw clockwise to increase the tension and counterclockwise to decrease the tension. The bobbin thread tension will not have to be adjusted very often if it is properly set. Some machines can go for years without the bobbin tension needing to be adjusted. For HA-1 style bobbin cases (the type pictured below), the bobbin case must be removed from the machine to adjust it.

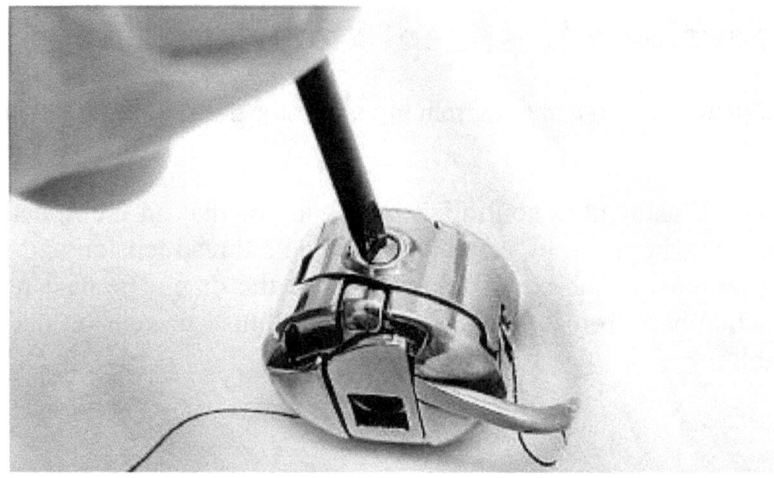

When adjusting the bobbin tension make small adjustments of 1/4th turn or less at a time, then test sew. Do it again if needed until the tension is correct. It does not take a large adjustment to make a noticeable difference in bobbin thread tension.

The machine pictured below does not have a user adjustable bobbin tension. See the section "Machines with no bobbin tension adjustment" at the end of this chapter if you think you have one of these machines or are not sure if your machine has a bobbin tension adjustment.

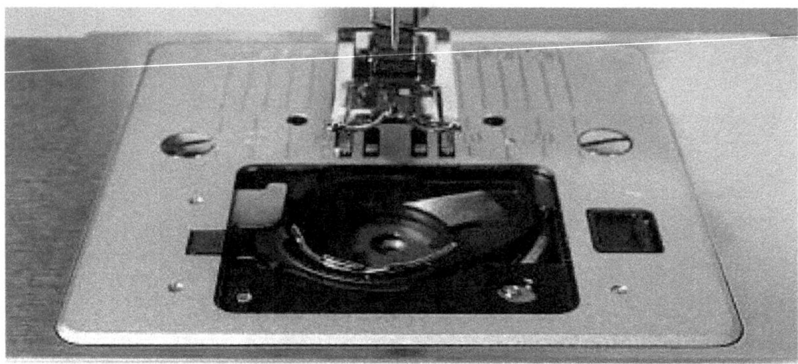

The goal is balanced tension

In the diagram below you are looking at a seam from the side as if the fabric was cut along the seam line. This is called a cross-sectional view. The fabric appears much thicker than real fabric is, this is done so that you can clearly see how the needle thread crosses the bobbin thread at the lock point. The tension is balanced when the lock point is close to the center of the fabric.

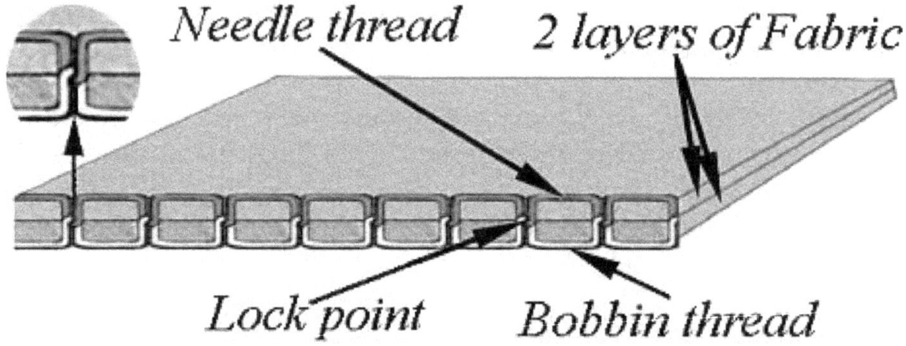

The next diagram shows what happens if the needle thread tension is too tight, correct or too loose.

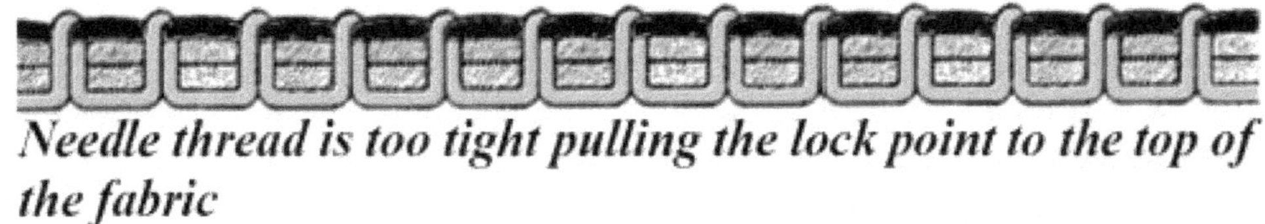

Needle thread is too tight pulling the lock point to the top of the fabric

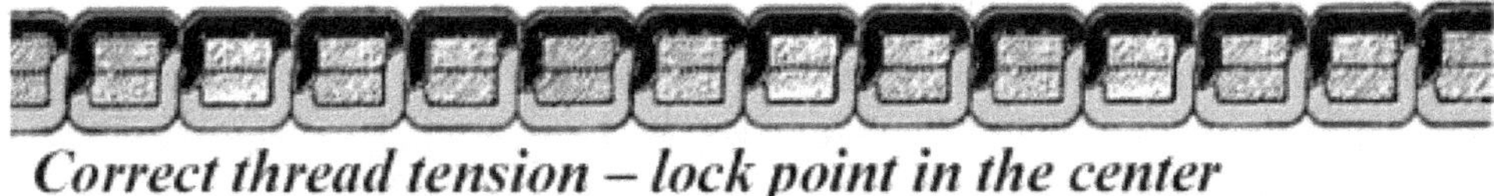

Correct thread tension – lock point in the center

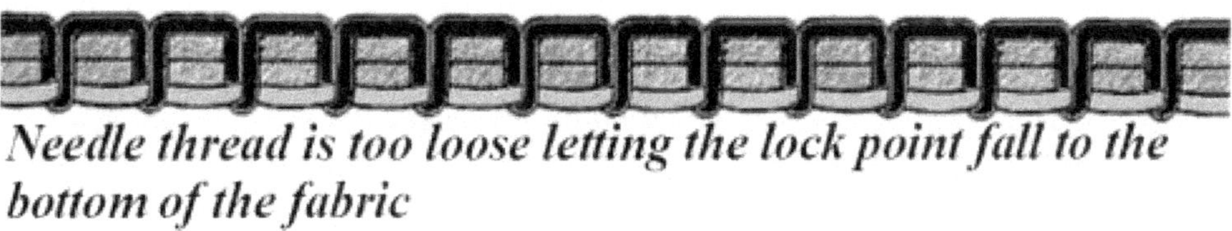

Needle thread is too loose letting the lock point fall to the bottom of the fabric

- If the needle thread is too tight (or the bobbin thread is too loose) then the needle thread will pull the bobbin thread up and the lock point will be on the top of the fabric (first picture). In this case the thread tensions are unbalanced.
- If the tension is balanced the lock point will be in the center of the fabric (second picture).
- If the needle thread is too loose (or the bobbin thread is too tight) then the bobbin thread will pull the needle thread down and the lock point will be on the bottom of the fabric (third picture). In this case the thread tensions are also unbalanced.

Recommended fabric and thread for testing

To test a machine for correct thread tension and stitch formation it is best to use opposite colors of thread and fabric so that the stitches are clearly visible. My preference is white fabric and black thread. Some people prefer to use different thread colors for the needle thread and bobbin thread so that it is easier to see the lock point (as shown in the next picture). For testing, it is recommended to use two layers of a medium weight cotton or polyester material (like a scrap from a bed sheet or pillow case).

Tension adjustment procedure

1. Thread the machine and set the stitch length to a medium-long setting. If you are using a zigzag machine set the zigzag width to zero (the straight stitch setting).
2. Set the needle thread tension to the default setting for your machine. This is the setting that the user manual recommends that you start with. For most machines that have numbered tension knobs or dials that go from 0 to 9 this will be a 3 or 4 on the dial (unless your user manual says otherwise, in that case follow your user manual). For machines that do not have a numbered dial then set the tension to about 1/4 to 1/3 of the maximum.
3. Sew a test seam for about 6 inches at a medium or fast speed and then remove the fabric from the machine and cut the thread tails. Keep track of which side of the fabric is the top side!
4. Inspect the stitches under a bright light. Use a magnifying glass or reading glasses if needed so that you can clearly see the lock point.
5. Is the lock point in the center of the fabric? If the lock point is in the center of the fabric then the tension is balanced.
6. If the lock point is on top of the fabric then loosen the needle thread tension or tighten the bobbin thread tension.

7. If the lock point is on the bottom of the fabric then tighten the needle thread tension or loosen the bobbin thread tension.
8. Most of the time you can just loosen or tighten the needle thread to get the lock point balanced. Most of your tension adjustments will be done with the needle thread tension knob or dial.
9. Once the tension is balanced, check to see if the overall tension is correct. If the stitches are loose or sloppy on both sides of the fabric then tighten both the needle thread tension and the bobbin thread tension slightly and test again.
10. If the stitches are too tight on both sides of the fabric then loosen both the needle thread tension and the bobbin thread tension and test again. See the pictures below in the section "Overall tension setting" later in this chapter.
11. After you loosen or tighten both the needle thread tension and bobbin thread tension to adjust the overall tension you should also recheck to see that the stitch is still balanced (lock point in the center of the fabric).
12. After you learn the basic tension check (steps 5 and 9) it will become very easy to check the tension by quickly looking at a seam. If you cannot achieve the correct tension your machine may have a problem or need adjustment. In this case see the chapter "Troubleshooting".
13. If you are using a zigzag machine then change to a zigzag stitch and set the stitch width to a medium or wide setting and go through steps 5 and 9 again and double check the tension. You may need to slightly increase or decrease the needle thread tension for zigzag stitches and then reset it when you go back to a straight stitch, this is normal. This will vary depending on the zigzag width, type of fabric and number of layers. If the tension is excessive the zigzag stitch will cause the fabric to fold and draw together in the center of the stitch parallel to the seam. For wide zigzag stitches you may need to loosen the needle thread, this will cause the stitch to become unbalanced. This is OK and it is not necessary to adjust the bobbin thread tension in this case to balance the stitch unless the bottom of the stitch will be visible. See the section "Lock point with long or wide stitches" in "Tension tips" later in this chapter.

Make sure to read the section "Tension tips" later in this chapter.

Quick bobbin tension adjustment for HA-1

If you have a machine with an HA-1 style bobbin-case (if your bobbin-case looks like the one in the picture below) there is a quick way to check the bobbin tension and do a rough adjustment (if needed) as follows:

1. Load and thread the bobbin-case with a full (but not over loaded) bobbin. Pull out about 5 inches of thread.
2. Hold the end of the thread and let the bobbin case hang by the thread in the air as in the picture below.
3. The weight of the bobbin case should not cause the thread to pull from the bobbin case. The bobbin-case should hang on its own without dropping down.
4. Jerk the thread lightly by quickly pulling up about 4 or 5 inches.
5. About 1 or 2 inches of thread should pull out of the bobbin case every time you jerk the thread.
6. If no thread pulls out then the bobbin tension is too tight. Turn the bobbin tension adjustment screw counter clockwise about 1/8th turn to 1/4th turn and try again. Repeat as needed until the tension is correct.
7. If more than 2 inches of thread pull out then the bobbin tension is too loose. Turn the bobbin tension adjustment screw clockwise about 1/8th turn to 1/4th turn and try again. Repeat as needed until the tension is correct.

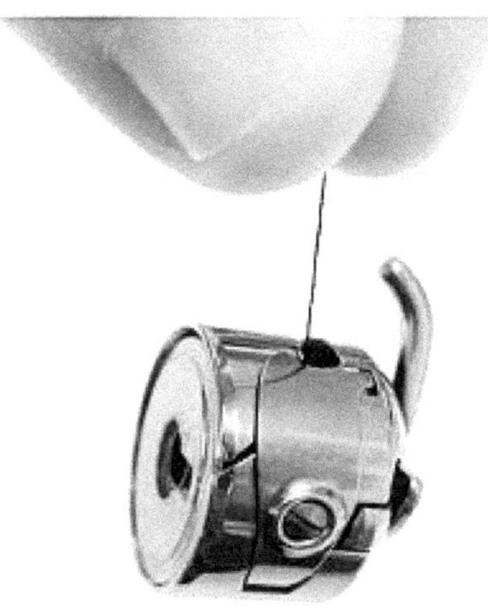

Overall tension setting

In addition to the tension needing to be balanced between the needle thread and bobbin thread, there must also be sufficient but not excessive overall tension on both threads. If there is insufficient tension then the stitches will be loose and sloppy. If there is excessive tension then the thread may break, the fabric may bunch up and curl or the machine may skip stitches.

Correct tension is the **LEAST AMOUNT OF TENSION** needed to form the stitch correctly, reliably and with the desired seam strength. There are several reasons for this:

- Your machine will run better and last longer with lower tension settings.
- It is easier to start with the thread tension too loose and adjust it tighter than to start with the tension too tight. This is because it is easy to see loose stitches and then tighten them until they are correct, but it is much more difficult to see if the tension is too tight.
- Excessive tension can cause a variety of problems such as weakening of the thread or fabric and poor seam strength.
- Loose tension also causes poor seam strength, but it is much easier to recognize loose tension.
- As a general rule when you are adjusting tension start with a lower setting than what you think will be needed and work your way up to the correct tension.

In the picture below the tension is correct. Notice that there is some curve to the stitches because the fabric in the picture has some springiness, if you are sewing with a flat, hard fabric there will be less curve at the top of the stitches. This tension setting is sufficient to bind the fabric for a strong seam.

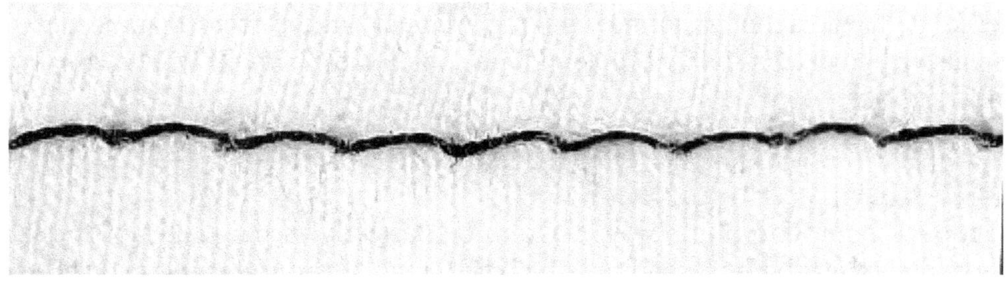

In the next picture the tension is too loose. Notice that the stitches are inconsistent and the thread is arched above the fabric on some stitches, this will cause a weak and sloppy seam. The thread will not bind the fabric and if the fabric is pulled hard during use, the loose stitches will allow the weave or knit of the fabric to separate causing failure at the seam.

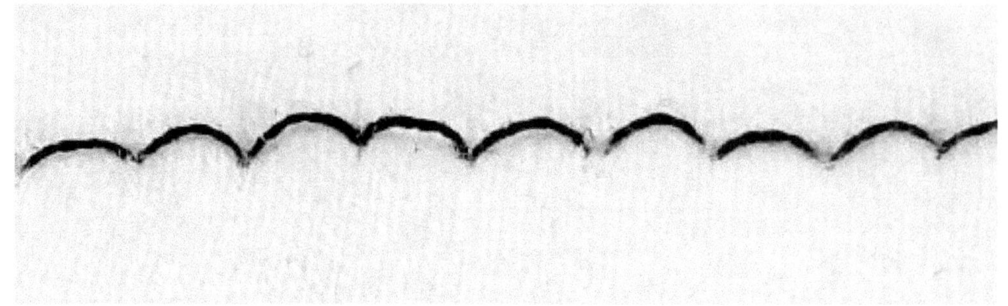

In the following picture the tension is too tight. Notice that the thread is pulled very tightly against the fabric and that the needle holes are enlarged by the excessive tension. Over time the thread will stretch and weaken causing the seam to be weak. In some cases the thread may even cut into the fabric causing fabric failure. Another problem with overly tight tension is that the seam can not stretch because the thread is already pulled tight and has no ability to stretch further. If the fabric is pulled and stretched during use, the thread in the seam will break.

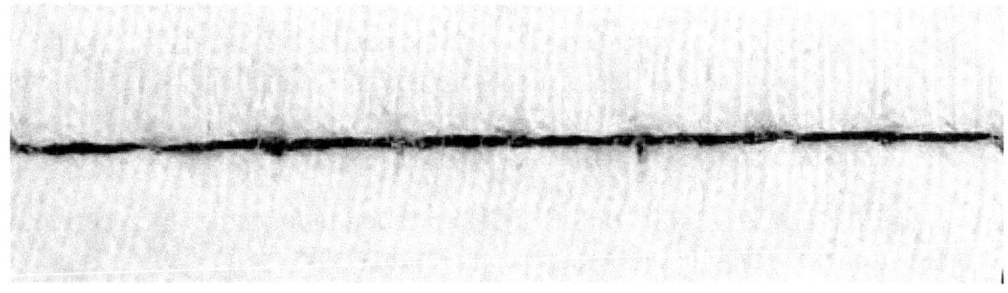

When sewing wide zigzag stitches or wide decorative stitches the tension must be reduced so that the fabric will not be drawn together parallel to the seam (as it is in the picture below).

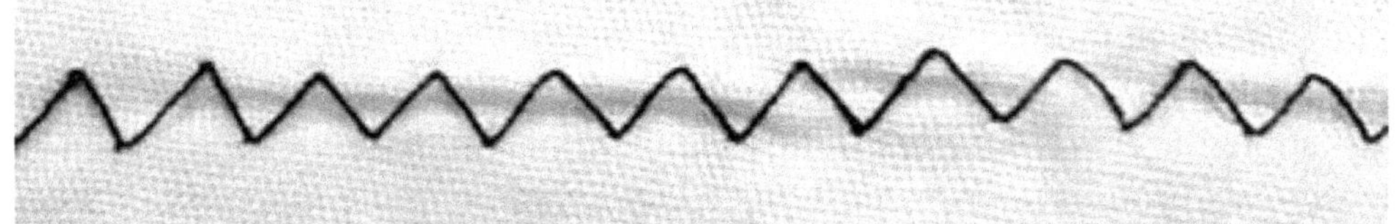

To prevent the fabric from being drawn together you can reduce the needle thread tension. In this case the stitch will be unbalanced, but this is OK unless you need to see both sides of the fabric because you will only be able to see the unbalanced stitches when looking at the bottom of the fabric.

For construction seams it is a good idea to always test the seam strength. Do this by pulling apart a seam and seeing if you can cause it to fail. If you find that more tension is needed so that the stitch will not pull apart then increase the tension enough to meet the requirement, but no more than needed. To gain stitch strength you can also try to decrease the stitch length, in some cases that will work.

Tension tips

Lock point - The lock point does not always have to be perfectly in the center of the fabric. Don't be overly obsessive and continually check and re-adjust the thread tension. Sewing machines are not prefect and the lock point will move towards the top or bottom of the fabric to some degree as you sew different thicknesses of fabric, as you adjust the stitch length or zigzag width. If the lock point is seriously wrong you should adjust it.

Decorative and wide zigzag stitches - Decorative stitches and wide zigzag stitches sometimes look better if the needle thread is slightly loose. This will cause the lock point to pull to the bottom side of the fabric. In seams where the bottom thread is not exposed this may give a better looking stitch.

Tension problems - If the tension is always changing and causing stitch problems then the machine may need troubleshooting and repair. After a while you will learn the difference between acceptable variations in the lock point and when there is a problem that requires troubleshooting and repair.

Observing tension as you sew - You should be aware of the stitches that the machine is making as you are sewing and occasionally check the tension and lock point (just by giving a quick look at the seam).

Quick check - If you are changing stitch lengths or widths, using decorative stitches, changing thread weights (heavier or lighter thread) or changing fabric you should check your tension (quickly do steps 3, 4, 5 and 6) to make sure that the thread tension is correct for the combination of stitch type, stitch length and width, fabric, and thread type that you are using.

Long stitches - You may have to reduce the needle thread tension when you use a long straight stitch and are sewing thin or light weight fabrics to avoid the fabric bunching. Bunching is when the stitch is tight enough to gather the fabric.

Wide stitches - When you use a wide zigzag stitch you may need to reduce the needle thread tension so that the fabric will not roll or bunch in the center of the stitch parallel with the seam. This happens mostly with thinner fabrics and happens most with a wide zigzag setting and a short stitch length setting (like a satin stitch).

Lock point with long or wide stitches - If you decrease the needle thread tension because you are using a wide zigzag stitch or long straight stitch with medium or thin fabric (to avoid bunching parallel to the seam) then the lock point will move to the bottom of the fabric (the stitch will be unbalanced because of the reduced needle thread tension). That is acceptable as long as the bottom of the fabric will not be visible in the finished item, you do not have to also readjust the bobbin tension every time you use a wide zigzag stitch.

Bobbin thread tension - The bobbin thread tension is typically not readjusted every time you change stitch lengths or widths, normally you adjust the bobbin thread tension only when you really need to. Some machines do not have a user adjustable bobbin thread tension.

Multiple bobbin-cases - If you have a machine with an easily removable bobbin-case such as an HA-1 style bobbin case or any of the newer Brother or Janome models that have a magnetically retained bobbin-case you may want to order a second bobbin-case for your machine. If you regularly sew with both heavy thread and regular thread you can have one case with the tension adjusted for the heavy thread and the other case adjusted for medium thread. In this situation you can just change bobbin-cases when you change threads instead of adjusting bobbin thread tension each time you change threads. Some bobbin-cases such as HA-1 bobbin-cases are very inexpensive.

Basting - Using decreased needle thread tension it is possible to make a basting stitch. To do this reduce the needle thread tension so that the needle thread is pulled all the way to the bottom of the fabric and bobbin thread is in a perfectly straight line on the bottom of the fabric. See the section "Basting" in the "Stitches" chapter for a picture of this and a better description.

Machines with no bobbin tension adjustment

Some newer machines with top-loading bobbins do not have user adjustable bobbin tension.

Does your machine have user adjustable bobbin thread tension?

- On most machines that are designed with user adjustable bobbin thread tension there will be a section in the owners manual that describes how to adjust the bobbin tension.
- If there is no section about adjusting the bobbin tension in your owners manual then your machine

probably does not have user adjustable bobbin tension. If you have a machine that does not have user adjustable bobbin tension then ignore the instructions for adjusting the bobbin thread tension and only follow the instructions for adjusting the needle thread tension.
- All machines with HA-1 style bobbin cases do have user adjustable bobbin thread tension even if it is not covered in the user manual.

On machines without user adjustable bobbin tension the bobbin tension is preset at the factory for general purpose thread. With some machines it is possible to find an adjustment screw and adjust the tension yourself anyway, even if the machine is not supposed to have user adjustable bobbin tension. This may violate the warranty if the machine has one. It is done as follows:

- Look on the bobbin-case and find the thread tension spring, this is the piece of metal that puts pressure on the thread to cause tension. Look for a small screw that goes through the tension spring to adjust it.
- The screw is usually locked with a drop of locking glue. This glue is called a "thread locker", but in this case the thread they are talking about is not sewing thread but the threads (spiral grooves) of the screw. "Thread locker" usually looks like yellow or dark red finger nail polish, but it could be another color. The thread locking glue (if there is any) can be partially removed by using a needle to carefully chip it away, then you turn the screw with a screwdriver until the screw is free to turn. Sometimes a solvent called thread lock remover is needed to soften the thread lock glue. Thread lock remover is similar to finger nail polish remover. Acetone may work as a remover for some types of thread locking glue.

On some machines there is no screw at all and if the bobbin tension is incorrect then the bobbin case must be replaced to correct the problem.

Problems???

If you have followed the procedures above and you are still having tension problems, perhaps your machine has a problem. Go to the chapter “Troubleshooting” and try to troubleshoot your machine or take your machine to a service shop for diagnosis and repair.

Stitches

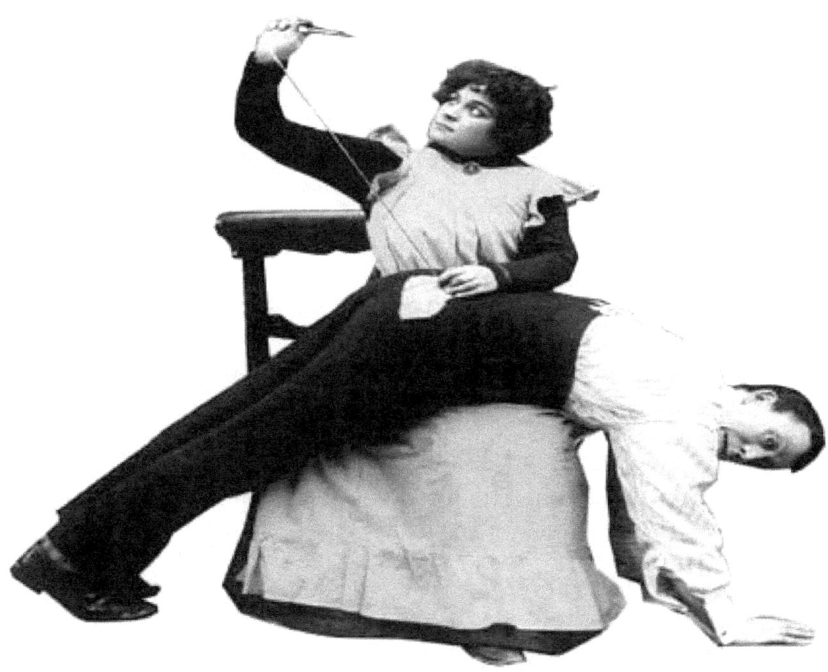

Basic stitch recommendations

For most fabrics - use a straight stitch with a medium stitch length of about 2 mm or 2.5 mm. Unless you have a reason to use a different stitch you can use this stitch for most of your sewing. If you have an older machine that uses Stitches Per Inch (SPI) instead of millimeters for the stitch length setting then you would need to use a stitch length of 12 SPI or 10 SPI. For more info on stitch length see the section "About stitch length" later in this chapter.

For stretch fabrics if you have a zigzag machine - Use a medium to wide zigzag stitch to prevent the thread from breaking when the fabric is stretched. The wider the zigzag stitch the more stretching the seam can handle. If the fabric is very stretchy use a shorter stitch length to allow for even more stretching of the seam. Always do a quick test by stretching the fabric to see if the seam will break before sewing a lot of fabric. If the seam breaks you need to use a wider zigzag stitch or a shorter stitch length.

For stretch fabric with a straight stitch machine - If you are sewing stretchy fabric with a straight stitch machine then reduce the stitch length and loosen up the needle thread tension a little so that the fabric can stretch without the thread breaking. Always do a quick test by stretching the fabric to see if the seam will break before sewing a lot of fabric. If the seam still breaks when you stretch the fabric and you are using the shortest stitch length possible then you will need to use a zigzag machine.

For decoration - use zigzag stitches or decorative stitches to jazz up your projects as needed. The decorative stitches you use are a matter of your artistic preference, try different stitches on a piece of test fabric until you discover what you like!

The owners manual for your machine will give you specific information on how to change the stitch length and what stitches your machine can make. Every model of machine is different so always read your owners manual.

Understanding stitches

When you look at a seam in a piece of fabric you are looking at the top view of the stitch, you are not seeing the entire stitch. The picture below shows a top view of a straight stitch.

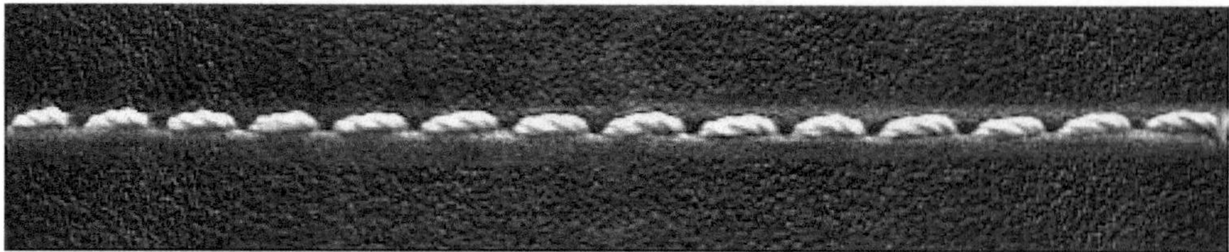

Straight Lockstitch

To see the entire stitch you would need to cut the fabric along the stitch line and then look at the stitch from the side. This is what a stitch diagram does, it is a cross sectional view (side view) of a stitch showing both the top thread and the bottom thread. Stitch diagrams let you visualize how different stitch types are made.

General purpose sewing machines (including all home sewing machines) make the lockstitch. In a lockstitch two threads are used to form the stitch. The threads interlace and form a "lock" in the center of the fabric.

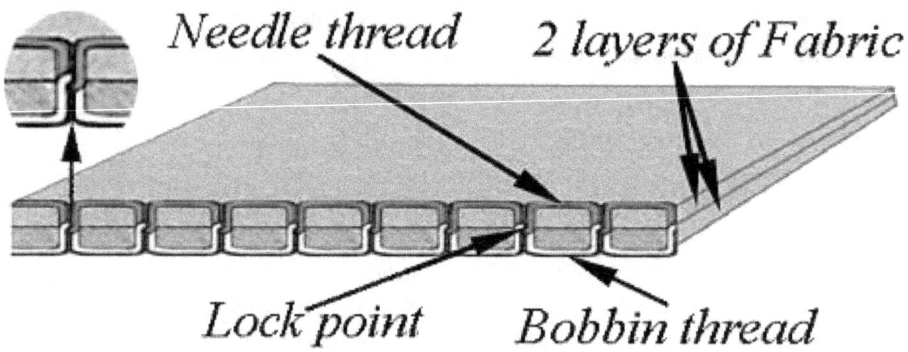

Stitch types

- **Straight** - The straight stitch is a lockstitch formed in a straight line.
- **Zigzag** - The zigzag stitch is a lockstitch formed when the needle is moved from side to side evenly on every stitch.

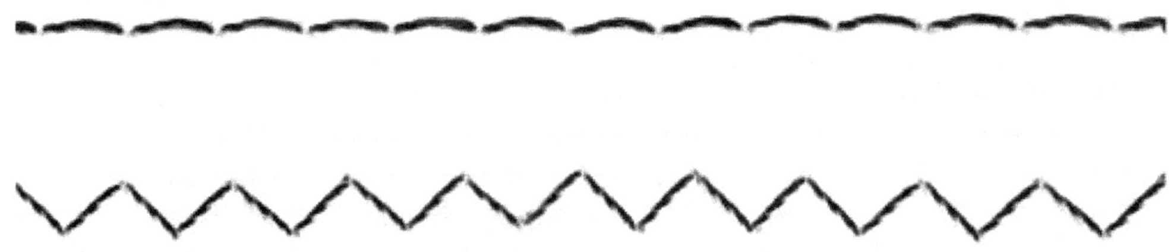

Straight stitch and zigzag stitch

- **Utility** – A utility stitch is any stitch type that is used for joining fabric. The most common utility stitches are the straight stitch and the zigzag stitch, but other stitches can be used as utility stitches. Most sewing is done with some form of utility stitch.
- **Decorative** – A stitch used for decoration. Most decorative stitches are based on the lockstitch. There are thousands of decorative stitches. To make decorative stitches the machine moves the needle from side to side according to a preset pattern while sewing. Some machines may also cycle the feed dogs into reverse while the machine is sewing according to the pattern to make more complicated decorative stitches.

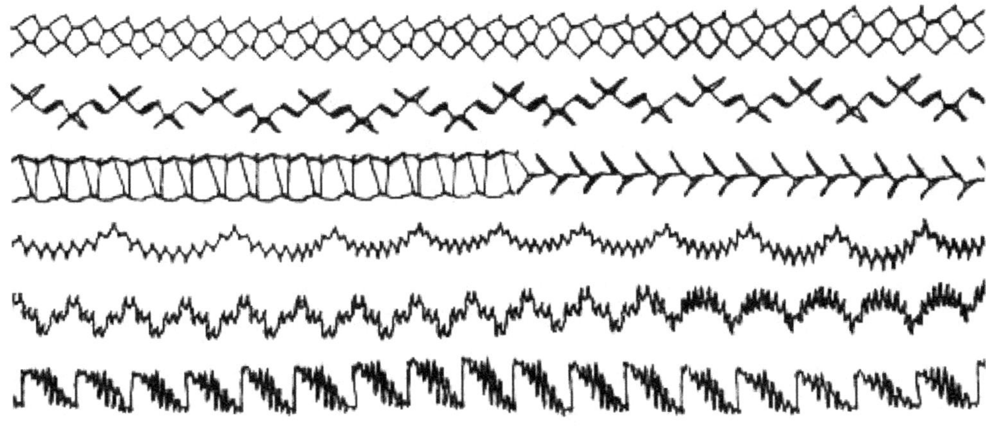

Decorative Lockstitch

About Stitch length

Stitch length is measured in Millimeters or SPI (stitches per inch):

Millimeters - This is the system used in newer machines and most older machines made in Asia. If the stitch length setting on your machine is numbered from 1 to 5 then (or 1 to 6 or 1 to 7) then your machine uses millimeters. To measure the stitch length measure 10 stitches of your seam and divide by 10. For example if 10 stitches measures 30 mm then 30/10=3 and that gives us a stitch length of 3 mm.

SPI (stitches per inch) - This is the system used in most older machines made by in the USA or made by Singer before about 1975. If the stitch length setting on your machine has numbers from 6 to 30 (or something in that area) then your machine uses SPI. The SPI numbers are inverse – longer stitch lengths have smaller numbers and shorter stitch lengths have larger numbers. For most general purpose machines a long stitch length is 6 or 8 SPI, a medium stitch length is 10 to 15 SPI and a short stitch length is 18 to 25 SPI. To measure the SPI you can hold a ruler parallel to the seam and count the number of stitches in a one inch long section.

Conversion - Sometimes you may need to convert from SPI to millimeters or from millimeters to SPI. You can use the following equations:

- Millimeters to SPI - To convert stitch length in millimeters to SPI the equation is 25.4/mm=SPI. In this equation 25.4 is the number of millimeters in one inch. For example if you wanted to convert a 2mm stitch length to SPI then the equation would be 25.4/2=12.7 so that is 12.7 SPI.
- SPI to Millimeters - To convert stitch length in SPI to millimeters the equation is 25.4/SPI=mm. For example if you wanted to convert a 6 SPI stitch length to millimeters the equation would be 25.4/6=4.23 so that is 4.23 mm.

Selecting the right stitch length

For many types of sewing, seam strength is not an issue and stitch length can be selected on the basis of aesthetics alone. Some things to keep in mind are that really short stitch lengths will use more thread and your machine will run much slower and take longer to complete your seams. Short stitches are also harder to remove if you need to take out the stitching. Really long stitch lengths may not look good and will not be as strong.

For most sewing a medium stitch length should be used such as 2mm or 2.5mm (12 SPI or 10 SPI). For stretchy fabric a shorter stitch length should be used or a zigzag stitch should be used to allow the stitch to stretch with the fabric, otherwise the thread may break.

Zigzag vs straight stitch

Zigzag Advantages and features:

- The zigzag stitch can be removed (ripped out) much more easily than a straight stitch for changes or alterations. This is because you can run a seam ripper straight down the seam cutting the cross threads of the zigzag stitch.
- The zigzag stitch can stretch. A straight stitch can be shortened to allow for a small amount of stretching, but usually it is best to use a zigzag stitch with stretchy fabric.

Straight Stitch advantages:

- A straight stitch is less visible and may look better than a zigzag stitch for some applications.
- A double straight stitch (two straight stitches closely paralleling each other) is generally stronger than a zigzag stitch unless fabric stretch is an issue, if the fabric will stretch then a zigzag stitch is the better choice.

Selecting the zigzag stitch width

The zigzag stitch can be used for strength or decoration. For decoration just use the stitch width that is the most aesthetically pleasing.

For added strength when sewing stretch materials the zigzag stitch allows the seam to stretch with the material instead of the thread breaking. The amount that the seam can stretch is dependent on the length and the width of the zigzag stitch, the shorter and wider the zigzag stitch the more the seam can stretch without breaking.

Basting

A basting stitch is a temporary stitch that can easily be removed. It is normally used to hold something in place. A handy basting stitch can be made by a lockstitch sewing machine. This is done by reducing the tension on the needle thread to prevent locking of the stitches in the center of the fabric. The bobbin thread will lie in a perfectly straight line on the bottom of the fabric allowing it to be pulled out when you want to remove the seam. If your needle thread tension dial setting is normally at 3 or 4 then try using a setting of around 1 or 1.5 for basting. In the picture below notice how the bobbin thread is tight and runs parallel to the bottom of the fabric and the needle thread is loopy and pulled all the way to the bottom of the fabric. Since the basting stitch is a temporary stitch it does not matter if it is a little loopy or not quite prefect.

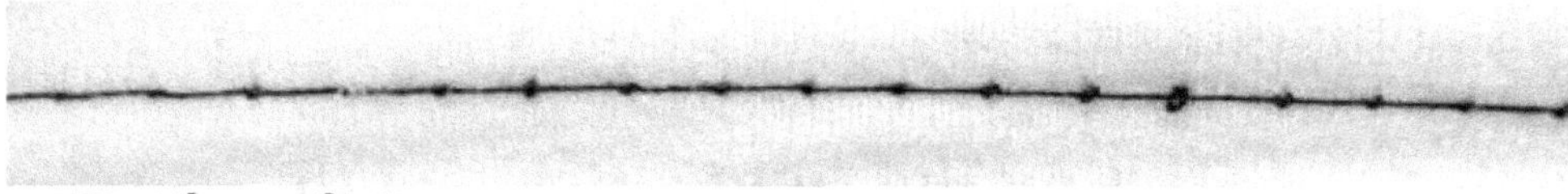

Basting stitch

When you want to remove the basting stitch you cut and then pull out the bobbin thread from the seam and then the needle thread will just pull free from the fabric. This is normally done 3 to 6 inches at a time.

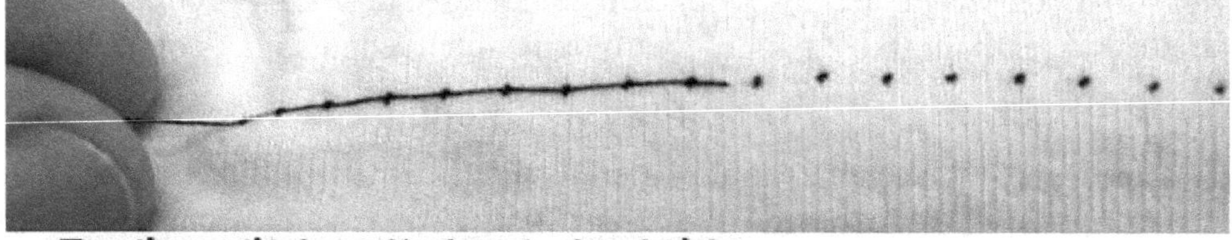

Basting stitch pulled out - backside

Thread

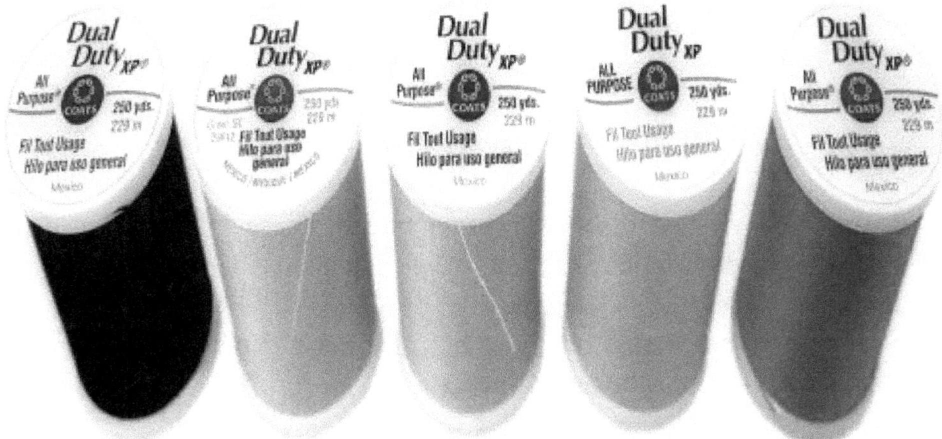

Thread basics

If you are doing general purpose sewing and don't want to spend time learning about thread, then just use a high quality polyester thread like one of the following:

- Coats & Clark - Dual Duty XP
- Coats & Clark - All Purpose
- Gutermann - Sew-All
- American & Efird - Excell

Modern general purpose polyester threads are so good that you can use them for most of your sewing and only look for other types of thread if you are doing specialized sewing that specifically requires a different thread. Sewing machines come from the factory adjusted for general purpose polyester thread. It is best to use a high quality premium thread because using cheap thread actually ends up costing more when you add up your time and problems.

About thread quality

What is the difference between a premium thread and a cheap thread? This depends on the thread, sometimes cheap threads can be OK, but there can be a lot of problems with them. Some of the considerations are as follows:

- Premium thread manufacturers have good quality control.
- Off-brand thread tends to be weaker then premium thread, if strength is a consideration you may have to use a heavier off-brand thread to get the same seam strength as a premium thread.

Just looking at thread does not give a good indication of quality, strange things can happen when the thread runs through a sewing machine such as shedding, stretching, knotting, untwisting or breakage. I have experienced thread that looks good and appears to work good at first, but later started shedding lint and the entire thread path of my machine became covered with lint after a short time.

How thread is made

Sewing thread is made by twisting many fibers together in a process called spinning. This is done in factories by large spinning machines, but can also be done using small hand powered machines. The fibers are first twisted into yarns (also called plies). These plies can't directly be used as sewing thread because they will always try to untwist and form knots and are therefore unstable. To counteract this untwisting effect a second operation is done in which several plies are twisted together in opposite directions to form a stable and balanced thread. Most general purpose threads use 3 plies.

Thread ply

- A number of strands are twisted together to form a thread. These strands are also known as the ply.
- A Three ply thread is made from three strands or plies that have been twisted together.
- Most general purpose sewing thread is three ply.
- Many times the ply is not stated on the label of thread. You can look at the thread under magnification to determine the number of plies.

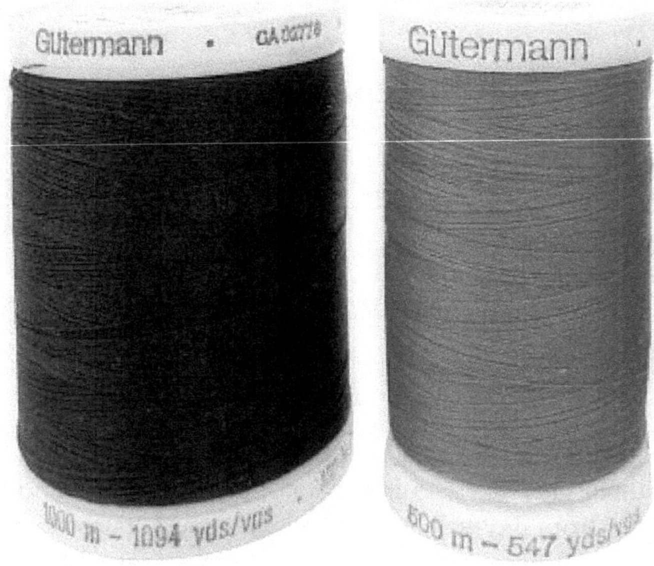

The picture above shows standard sized spools of thread used for home sewing machines, the spool on the right holds 500 meters and the spool on the left holds 1000 meters.

Thread Size Systems

Threads come in a bewildering array of types and sizes. To make matters worse there are several non-compatible thread size systems and each supplier seems to use a different system.

- Before sewing machines there were only a few sizes of thread, such as light, medium and heavy. As sewing machines and weaving machines were developed for specialized purposes, thread was developed for these machines in numerous sizes and types. This development happened separately in many countries around the world and consequently many different thread size systems were developed.
- Most threads sold for home use are not labeled with a thread size. You have to search on the suppliers web site for the thread size information and possibly download a specification sheet.

The Tex system

- The Tex system is the most widely used thread size measurement system and is used worldwide.
- Most thread suppliers will list the Tex sizes for their threads or have a conversion chart available so that you can convert their thread sizes to Tex.
- The Tex system is intuitive because the numbers increase as the thread sizes get larger. Some other systems have smaller numbers for larger thread sizes or the numbers are arbitrary and this is confusing.
- The Tex size is determined by the weight (in grams) for 1000 meters of thread. The Tex system is called a fixed length system because a fixed length of thread (1000 meters) is weighed to determine the thread size.
- The common Tex sizes that can be used with home sewing machines are 21, 24, 27, 34, 40, 50, 60, 70. To give you an idea of what these sizes are, TEX 21 is a very light thread, TEX 34 is a normal general purpose sewing thread and TEX 70 is a heavy upholstery thread.
- Tex size will appear on the thread spool as T-nn (such as T-40 or T-70 or whatever the size is, nn means number).

Other size systems

Denier (d) is another fixed length system similar to Tex except that the Denier size is the weight in grams for 9000 meters of sewing thread. The Denier size is 9 times the Tex Size. Denier sizes will appear as nnd (such as 55d or 70d or whatever the size is). To convert from Denier to Tex the equation is den/9 = Tex

Commercial Size - The Commercial Size system is for threads used in upholstery, technical sewing (camping gear, tents, backpacks), webbing and straps for trucks and boats, sails, etc. Common sizes are 15, 23, 33, 46, 69, 92, 138, 207, 277, 346, 415 and 554. The Commercial Size system is based on the Denier system divided by 10. To convert from Commercial Size to Tex the equation is (Size*10)/9 = Tex. For example Commercial Size 69 converted to Tex (69*10)/9 = 76 and in Tex we round that down to the nearest step size of T70

Metric Counting (Nm) - Nm stands for "Number Metric". Nm is the number of meters of a strand that weighs one gram followed by the number of strands that make up the thread. For example Nm 120/1 is composed of one strand of thread and one gram of that strand is 120 meters long. 50/3 is composed of three strands of thread and one gram of a single strand is 50 meters long. The Metric Counting system has mostly been replaced by Label Number (No).

Cotton Count (Nec), (Ne) or (cc) - Used for all types of spun threads including polyester thread. The name Cotton Count originated before polyester came into use. Cotton Count is the number of strands that are 840 yards long that it takes to equal 1 pound. A Cotton Count of 20 means that 20 strands of thread 840 yards long weigh 1 pound. To convert from Cotton Count to Tex the equation is 590.54/NeC * Strands = Tex. In this case "Strands" equals the number of strands in the thread, for example if you wanted to convert a Nec 50/3 to Tex the equation would be 590/50 * 3 = 35.4 and we just round that down and call it Tex 35

Thread size recommendations

- Lightweight fabrics T-24
- General purpose home sewing T-34
- General purpose home serger T-27
- Jeans and denim T-50 to T70
- Window coverings T-34
- Upholstery – Light to Medium T-34 to T-50
- Upholstery – heavy T-70
- Car upholstery T-50 or T-70

- Backpacks and outdoor gear T-50 or T-70
- Soft leather T-40 to T-70

These are general recommendations, you will have to test for strength and appearance and change if needed for your specific application. If you are sewing specialized items like sailboat sails it is a good idea to look at the various forums on the Internet, they often talk about specific recommendations for thread size. If you are sewing on a home sewing machine you will be limited to the maximum thread size that the machine can handle.

The following are general recommendations for the maximum thread size capability of home machines:

- Most newer home machines can use up to T50.
- Most newer heavy-duty machines can use up to T70, but run better with T50 or T60.
- Old heavy-duty straight stitch machines like Japanese HA-1 or Singer 15-91 machines can use up to T70 with no problems.
- Most old Japanese cast iron zigzag machines can use up to T70, but run better with T50 or T60.

Thread types

Polyester

- Polyester is the best all purpose sewing thread and should be your first choice unless you have a clear cut reason to use another type of thread.
- Polyester thread is the most used of all sewing threads because it has the best performance for general sewing.
- Made from synthetic fibers (thermoplastic).
- High strength.
- Polyester stretches well and can stretch up to 25 percent before breaking. This is good for sewing stretch fabrics.
- Resistance to heat is OK, it gets sticky at 440°F and melts at 475°F
- Resistance to abrasion is very good, but nylon is slightly better.
- Resistance to chemicals is excellent.
- Color fastness is good.
- UV resistance is very good. Polyester is preferred over nylon for outdoor applications with long term sun exposure.
- Low lint, does not shed very much when going through sewing machines.
- Relatively low cost.

Cotton

- Cotton is preferred for decorative stitching and quilting applications that require vibrant colors.
- Preferred for sewing cotton garments that can be ironed at high temperatures.
- Made from natural plant based fiber.
- Takes dyes well and can have vibrant colors.
- Cotton has Low strength compared to polyester or nylon.
- Does not stretch very much. Not good for sewing stretch fabrics because it can break.
- Good Heat resistance and can take higher temperatures than polyester of nylon. Does not melt when ironed at high temperatures.
- Resistance to chemicals is not as good as polyester or nylon.
- Resistance to abrasion is not as good as polyester or nylon.
- Color-fastness is not as good as polyester.

- Cotton thread can be mercerized. This is a chemical process that adds luster and strength to the thread. Mercerized cotton thread has an appearance somewhat similar to rayon. Mercerized thread tends to shed more lint.

Rayon

- Rayon is preferred for embroidery and decorative applications that require the most vibrant colors.
- Rayon is a semi-synthetic fiber that is made from highly processed natural fiber.
- Characterized by very vibrant colors, high sheen and deep luster.
- Medium to low strength. Not as strong as polyester or nylon.
- Does not stretch very much. Not good for sewing stretch fabrics because it can break.
- Resistance to chemicals is not as good as polyester or nylon.
- Resistance to abrasion is not as good as polyester or nylon.
- Color-fastness is not as good as polyester.

Nylon

- Nylon thread is good for backpacks, luggage, cases, shoes, upholstery and other applications that require high strength and abrasion resistance.
- Not good for permanent sun exposure such as car upholstery, awnings or sunbrellas, UV resistance is OK but not good. For better long term UV resistance for applications with continuous sun exposure use a polyester thread of a similar size.
- Made from synthetic fibers (thermoplastic).
- High strength.
- Nylon stretches well and can stretch up to 25 percent before breaking.
- Resistance to chemicals is good.
- Resistance to abrasion is very good.
- Resistance to heat is OK, it gets sticky at 440°F and melts at 495°F
- Color fastness is good.
- UV resistance is good but not great. Polyester is preferred over nylon for outdoor applications with long term sun exposure.
- Almost no lint, does not shed when going through sewing machines.
- Bonded nylon thread has a hard coating that increases abrasion resistance and makes the thread feed better through the sewing machines.
- Relatively low cost.

Mono-filament

- Semi-transparent and is almost invisible. This is good for applications where the seam should blend in and not be seen.
- Made from nylon or polyester and looks similar to fishing line, but is much softer and will feed properly through a sewing machine whereas fishing line will not feed properly.

Strength testing

Basic testing of the breaking strength for smaller size threads can be done by breaking various threads you are testing with your hands. This works well if you only want to get a general idea if one thread is stronger than another. It is best to wrap each end of the thread around a pencil or pen leaving about a foot of thread between the two pencils and then pull the pencils apart until the thread breaks.

Needles

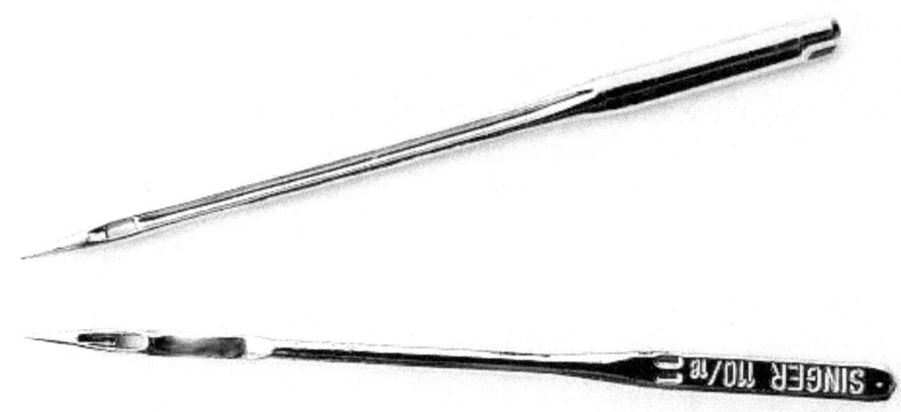

Needles for home sewing

For most home sewing with a general purpose polyester thread just use a size 14 universal needle. If you are sewing thick fabric (Jackets or jeans) use a size 16 universal or denim needle. If you are sewing very thin fabrics you can use a size 12 needle.

Change your needles after about 8 hours of continuous sewing or when the needle becomes blunt of damaged. To learn how to check your needle see the section later in this chapter about Needle Inspection.

Buying needles

When buying needles for your machine you need to know the following:

- **The Size** – size 14 for example. See the next section for more about needle size.
- **The System** - The needle system will be a strange looking number like 130/705H for example. Each needle company has their own part number for home needles; the Schmetz part number is 130/705H, the Organ part number is HAx1, the Singer part number is Style 2020 and the generic part number is 15x1H. They are all the same and will work in any modern home sewing machine.
- **The Type** – The needle type specifies the intended use such as "Jeans/Denim", "Leather" or "General Purpose". Many times a type number will added to the system number as a suffix for example "-J" for jeans. See the section "Needle Types" later in this chapter for a list of many popular needle types.

When you know the Size, System and Type you have the required information to buy your needles, for example "size 14, 130/705H-J".

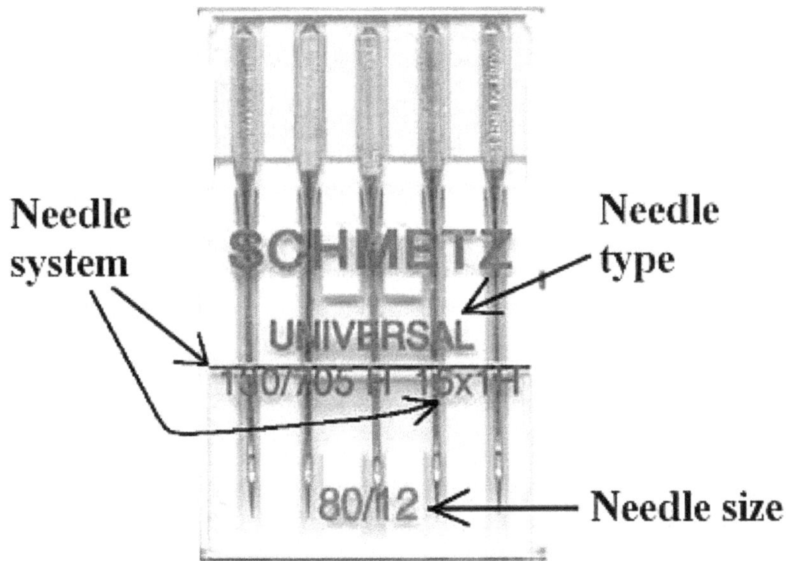

Some needle manufacturers will list more then one system number on their packages. In the picture above two needle system numbers are on the package. The first one is the companies system number and the second one is a generic system number that means the same thing.

Needle size

The size of the needle is determined by the diameter of the shaft.

There are two sizing systems, the Singer system and the metric system:

- The Singer system is also sometimes called the American system. The sizes in this system are arbitrary, but the size does get larger as the needle diameter increases. For home sewing machines the sizes range from 9 to 18.
- The Metric system (called NM or "number metric") is the actual measurement of the blade diameter in hundredths of a millimeter. For example a 100 needle has a diameter of 1 millimeter. As the needle diameter increases the size gets larger.

In many cases both sizes are used. For example a needle that is labeled 14/90 or 14-(90). The first number 14 is the Singer size and the second number (90) is the metric size. Some manufactures list the metric size first and the Singer size second with a forward slash / between the sizes (such as 80/12). If only the Singer size is given the package will read "size 14" and if only the metric size if given the package will read "NM 90".

The most common needle sizes are: 9-(65), 10-(70), 11-(75), 12-(80), 14-(90), 16-(100), 18-(110).

Selecting the right size needle and thread

Use the charts that follow to get a rough idea of needle and thread sizes for common types of fabrics. Also look to see what needle and thread experienced people are using to sew similar items.

The number of layers of fabric and total seam thickness should be taken into account. When many layers of fabric or thick folded seams are sewn, a larger size needle must be used for greater strength to prevent bending and breaking of the needle. For example if you use a size 14 needle and Tex 34 thread for sewing two layers of denim fabric, then you may need a size 16 needle and Tex 40 thread for 6 layers of denim. You also should take into account the layers that are folded into the seam on seams that are folded over such as in the crotch or zipper areas of pants.

Needle sizes and recommended thread sizes

Singer Needle Size	Metric Needle Size	Spun Polyester Thread Size	Commercial bonded Nylon or Polyester Thread Size
9	65	T-18	15
10	70	T-24	15
11	75	T-27	15
12	80	T-34	33
14	90	T34-T-45	33-46
16	100	T45-T-60	46-69
18	110	T-70	69

Needle sizes for common fabric types

The recommendations in the following chart are general, the size of thread can vary quite a bit from manufacturer to manufacturer, so it is a good idea to use the quick needle size test in the next section to be on the safe side.

Fabric Type	Thread Size	Needle Size
Lightweight fabrics	T-24	10-11
General home sewing	T-34	12-14
General home serger	T-27	11-14
Heavy fabric in home serger	T-34	14
Jeans and denim	T-50 to T-70	16-18
Window coverings	T-34	12-14
Upholstery – Light & Medium	T-34 to T-50	14-16
Upholstery – heavy	T-70	16-18
Car upholstery	T-70	16-18
Backpacks and outdoor gear	T-70	16-18
Soft leather	T-40 to T-70	14-18

Quick needle size test

A quick test to see if a needle size will work with a certain thread size:

1. Remove the needle from the machine (or use a spare needle of the same size and type). Cut off a few feet of the thread you want to test and thread the needle with it.

2. Hold the ends of the thread in the fingers of your opposed hands, so that the thread is parallel to the ground between your two hands. Pull the thread somewhat tightly (not enough to stretch the thread) between your hands with the needle about half way between your hands.

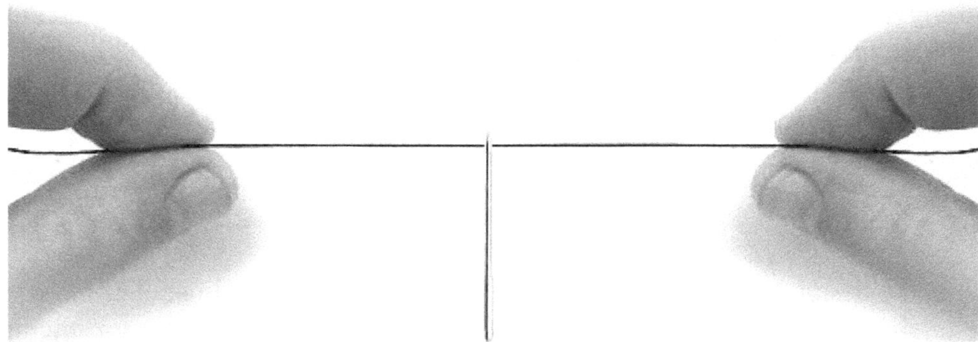

3. Move one of your hands up and the other down, so that the thread is tilted at a 45 degree angle to the ground and let the needle slide down the thread until it runs into your fingers.

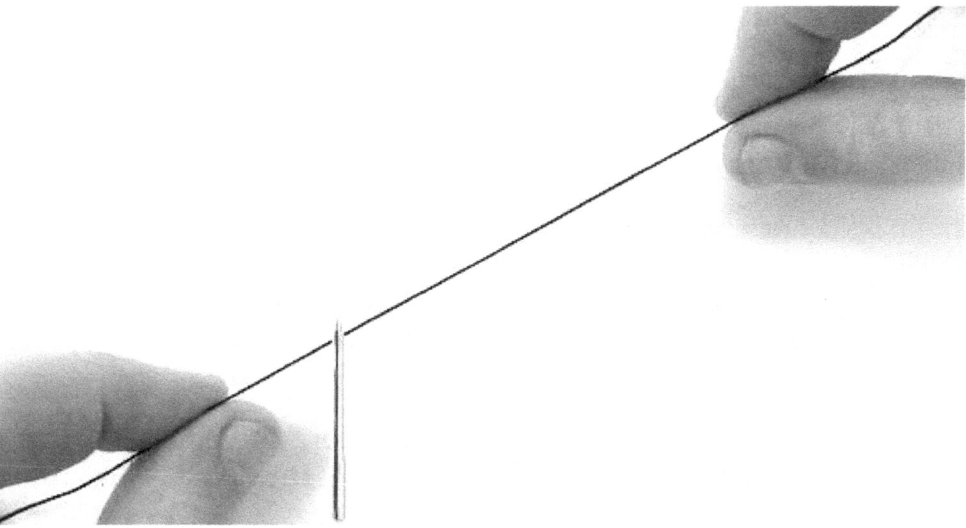

Then reverse your hands (lower the upper hand and raise the lower hand) so that the thread is tilted 45 degrees in the other direction and the needle slides down to your other hand. Do this several times, so that the needle slides back and forth on the thread.

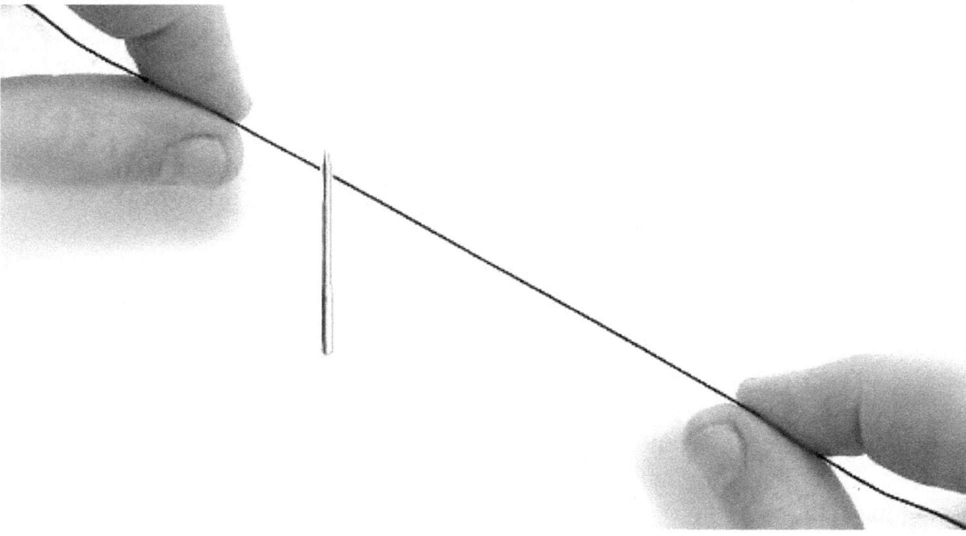

4. If the needle does not slide freely the needle is too small. The needle should slide smoothly and not catch on the thread or slide erratically.

5. Now look closely at the thread going through the eye of the needle. The thread should almost touch the sides of the eye of the needle. If the eye is oval the thread does not need to be close to the top and bottom of the eye, just the sides of the eye. If the thread is swimming in the eye (lots of extra space) then you may want to use a smaller size needle.

Needle types

The needle type is the designation for the kind of tip, eye and the intended use. For example a Topstitch needle has a semi-sharp tip and a large oval eye and is intended to be used for topstitching. A Leather needle has a chisel tip and a medium size eye and is intended for sewing leather and non woven materials.

The following list covers the most common needle types (such as regular, ball point, etc).

Universal, Regular or General Purpose – All mean the same thing. The tip is sharp but slightly rounded, so that this type of needle is optimized for woven fabrics, but will also work with knit fabrics. This is the most commonly used of all needle types and the needle type that comes with new sewing machines. Some people use a general purpose needle for sewing everything. For a heavier duty needle that is similar to the general purpose needle see the Quilting needle below. Different needle manufacturers have their own designations for the general purpose needle type, for home sewing machines the types are 130/705 H (Schmetz), HAx1 (Organ), 15X1 (many suppliers), Style 2020 (Singer).

Ball Point – The tip is very rounded (like a ball) and is made for sewing stretchy knits such as synthetics. The rounded tip allows the needle to push the threads of knit to one side and pass between them. A sharp tip needle would try to go through the threads instead of going around them and would cut or break threads causing fabric damage and erratic stitching. For home sewing machines the types are 130/705 H-S (Schmetz), HAx1SP (Organ), Style 2045 (Singer).

Microtex/Microfiber – Very sharp needle for sewing microfiber, silk and other tightly woven fabrics. For home sewing machines the types are 130/705 H-M (Schmetz), HAx130SPI (Organ).

Denim/Jeans – For sewing jeans and canvas. This needle has a sharp tip and reinforced blade for going through thick fabrics and seams. For home sewing machines the types are 130/705 H-J (Schmetz), HAx1DE (Organ), Style 2026 (Singer).

Leather – Chisel tip that cuts through the material making a good quality stitch on leather or vinyl. It should only be used for real leather or vinyl, for synthetic leather (polyurethane), pleather or faux leather a denim needle or quilting needle should be used. For home sewing machines the types are 130/705 H LL (Schmetz), HAx1LL (Organ), Style 2032 (Singer).

Embroidery – Large eye for passing decorative threads and the tip is semi-sharp. For home sewing machines the types are 130/705 H-E (Schmetz), HAx1EB (Organ), Style 2000 (Singer).

Quilting – Semi-sharp tip and reinforced blade to limit bending caused by the fabric motion during free-motion quilting. This is the most rugged needle available for home sewing machines and makes a good general purpose needle for heavy use. For home sewing machines the types are 130/705 H-Q (Schmetz), HLX5 (Organ).

Topstitch – Large oval shaped eye and large grooves to accommodate thicker thread sizes. Topstitch needles are also used for embroidery when a larger eye is required. The tip is semi-sharp. Some people use the topstitch needle with two threads instead of a single thicker thread. For home sewing machines the types are 130 N (Schmetz), 130N (Organ).

Metallic – Large eye and grooves, semi-sharp tip. This needle is for decorative metallic thread types and is similar to the topstitch needle which can be used for the same purpose. For home sewing machines the types are 130 MET (Schmetz).

Wing/Hemstitch - The wing needle has a blade that is shaped like cobra head or a wing that extends to the sides. This wing makes a larger hole in the fabric and is used for heirloom sewing with fancy pattern stitches. In many fancy pattern stitches the feed dogs reverse during stitch formation and the needle must enter the same hole more than one time. The larger hole produced by the wing needle allows the needle to re-enter the same hole more accurately and produces a better looking stitch. For home sewing machines the types are 130/705 H WING (Schmetz), Style 2040 (Singer).

Twin – The twin needle has two needles mounted to the same shank. It is used for topstitching or decorative stitching. There are several models with spacing between the needles from 1.5mm to 6mm. NOTE: Only some zigzag machines can use twin needles, many machines can not use twin needles including machines with a side-facing vertical hook and all straight stitch machines. Read the user manual for your machine to make sure that your machine can use a twin needle and to find out about any specific settings needed for twin needles. Twin needles for home sewing machines the types are 130/705 H ZWI (Schmetz), Style 2025 (Singer). Schmetz also makes many other types of twin needles such as Jeans Twin 130/705 H-J ZWI and Embroidery Twin 130/705 H-E ZWI.

Triple – Same as a twin, but has three needles mounted to the same shank.

Embroidery Spring Needle – The spring needle is used for free-motion embroidering with a hoop or movable frame. This needle has a spring mounted to the shaft of the needle and is made to be used with no presser foot (the presser foot must be removed from the machine). For home sewing machines the type is 130/705 H SPR (Schmetz).

Parts of the modern needle

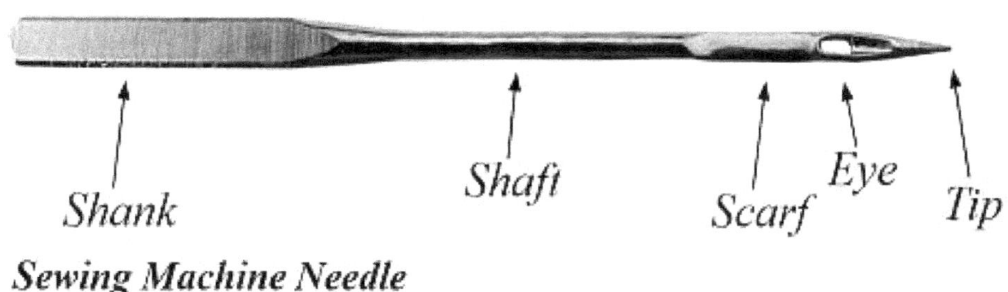

Sewing Machine Needle

Shank - The shank of the needle is the part that goes into the needle bar of the machine and is held in place by a needle clamp or set screw. Home needles have a round shank with a flat section so that they can only go into the machine properly in the correct orientation.

Shaft - The shaft of the needle is the area between the shank and the scarf.

Scarf - The scarf is smooth flattened area in the shaft of the needle just above the eye of the needle.

Eye - This is the part of the needle with a hole for the thread to pass through. On larger needle sizes the eye will be larger to handle larger thread sizes.

Point - This is the part of the needle starting from the eye and tapering to the tip.

Tip - This is the part of the needle that first penetrates the fabric.

Inspecting needles

Visual inspection - To inspect the tip of a needle you will need a high power magnifier or loupe, a magnification of X10 or greater is best. Inspect under a bright source of light. Remove the needle from the machine and compare the tip with a new needle. If you can see any wear or damage then replace the needle. Pay close attention to the tip of the needle, the tip can become chipped or damaged easily. If you see wear around the eye of the needle your machine may be out of adjustment.

Checking the tip by feel - Some people can fairly accurately tell if the tip of needle is in good condition by feeling the point of the needle with the tips of their fingers. This is best done by comparing the needle with a new one. To find out if you are good at this try to feel the tip and then remove the needle from the machine and do a visual inspection to confirm if you are correct with your assessment. One advantage of checking the tip by feel is that you can do it with without removing the needle from the machine. As a quick check, even if you can't tell if the needle is perfect or not, you should be able to tell if the needle is really worn or damaged.

Checking needles for straightness - To check needles for straightness you will need a perfectly flat surface. A piece of window glass works well for this, you can get a small square of window glass from a hardware store or a glass store. Have them sand down the edges so that you do not cut yourself. In the picture below a flat piece of stainless steel is used. Always compare the needle you are checking with a new needle. You may also want to check new needles before using them. I have come across batches of new needles that were defective (slightly bent), even from major manufacturers.

When checking home needles press the flat side of the shank against the flat surface of the glass and look at how far the tip of the needle is from the glass surface. Compare this with the new needle. If you are not sure that your new needles are perfect then compare several new needles from different manufacturers. If the distance of the tip to the flat surface is different for the needle that you are checking then it is bent or defective.

The first picture below shows a perfectly straight needle, notice how the shaft of the needle is completely parallel to the surface of the metal.

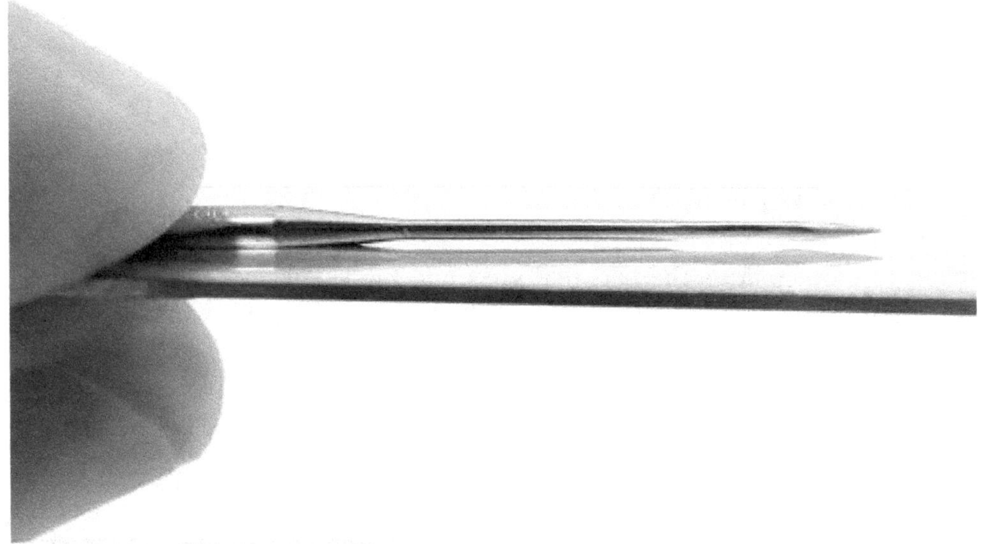

This needle is straight

The second picture shows a bent needle, in this case it is bent up and away from the metal surface. This is actually bent more than you would normally see with a bent needle, I wanted to be sure you could see it in the picture. If your needle is not completely parallel to the surface then it is bent and should not be used. If you are in doubt then compare several needles from different manufactures to get an idea of what is normal variation from needle to needle and what is clearly a bent needle.

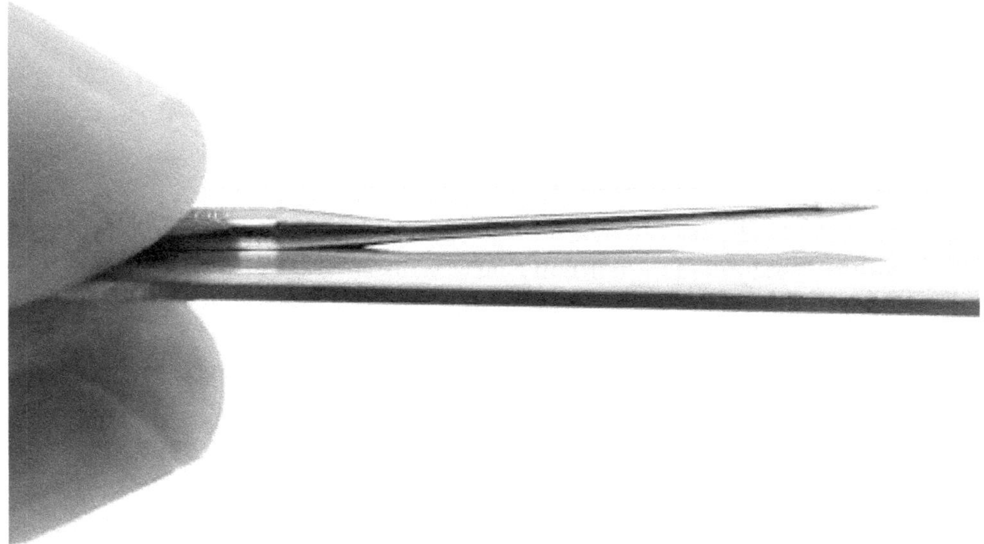

This needle is bent up

In the picture below the needle is bent down and is actually touching the metal surface.

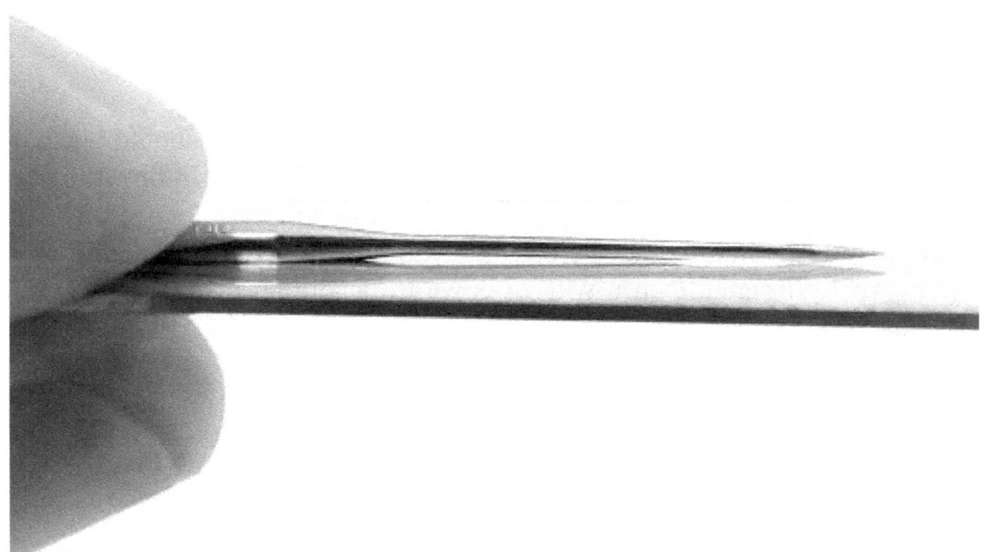

This needle is bent down

How long do needles last?

There is a huge variation in how long sewing machine needles last depending on many factors:

- **Damage** - Any time a needle becomes damaged its life is over. I needle can become damaged in the first few seconds if it comes in contact with the needle plate or other metal parts of the machine. This can happen if the fabric is pulled while sewing or if the machine is running over heavy seams and the needle is deflected while going through the fabric as can sometimes happen. Also the needle can become damaged if the machine is out of adjustment causing the needle to contact metal parts of the machine. If a needle is not damaged it can last a long time.
- **Wear** - As the needle is used it will wear due to contact with the fabric and possible contact with metal parts of the machine. Each type of fabric has its own level of abrasiveness that effects how the needle wears. The adjustment of the machine effects how the needle wears.

- **Needle quality** - A high quality needle made from high grade metals with properly hardened and polished surfaces can last many times longer than a poor quality needle.

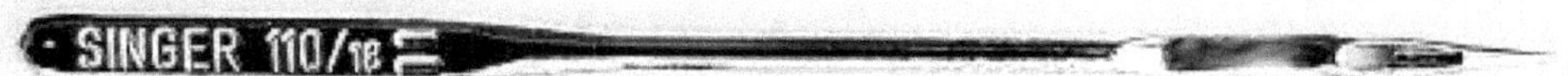

There is no set time limit for needle life. It is best to periodically inspect needles. With inspection you will get an idea of the length of time a needle normally lasts with your specific machine and type of sewing, but keep in mind that a needle can get damaged at any time.

Some sewing machine manufactures (and sewing books) recommend that you change needles after a set number of hours of use (like 8 hours of use). This recommendation is made with the assumption that you don't know how to inspect or check your needles. I make the same recommendation for home sewers at the start of this chapter to be on the safe side.

Installing needles

When changing the needle make sure that the needle is seated properly in the needle bar before tightening the needle clamp.

Some machines have an overly tight or loose fit of the needle shank to the needle bar.

- If the fit is loose then the needle may tilt from side to side, so you should check to see that the needle lines up to the center of the needle hole before you fully tighten down the needle clamp. If the needle is off to one side it can cause problems or needle breakage.
- If the needle fits tightly into the needle bar then you have to make sure that it is inserted all the way, if it is partially inserted then it will go down to far during operation and could contact the hook mechanism and break.

Bobbins

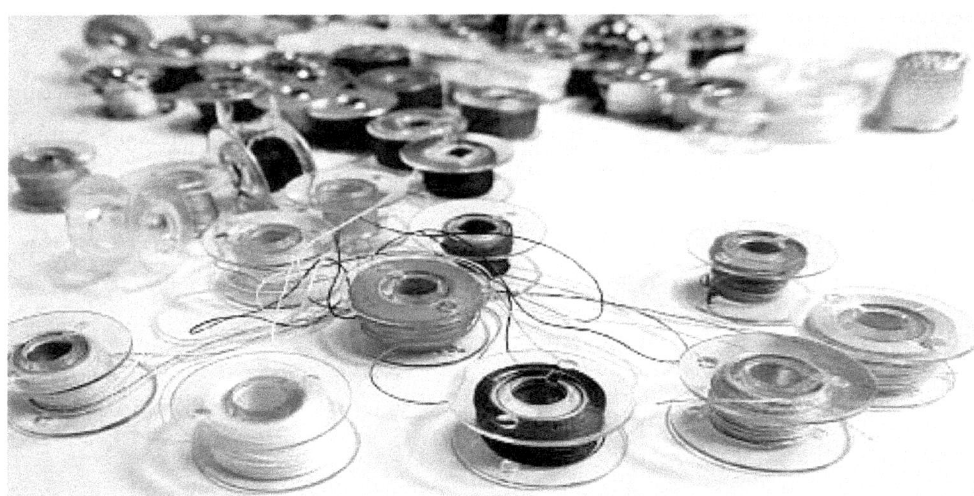

Bobbin basics

A bobbin is a small spool that is used in a lockstitch sewing machine to hold the lower thread. The bobbin is inserted inside a bobbin-case.

When troubleshooting a sewing machine it is easy to overlook the simple bobbin. What could possibly go wrong with a bobbin? Actually bobbins can cause some of the nastiest tension problems. The bobbin must turn freely or you will have tension problems, erratic stitching or thread breakage.

Unfortunately there are quite a few bobbins that are defective from the factory, these bobbins are warped or have imperfections that interfere with the bobbin turning smoothly in the bobbin-case. There are some suppliers that sell cheap bobbins of bad quality.

Make sure you are using the correct bobbin for your machine, the wrong bobbin will not turn smoothly. Metal bobbins have been made for over a hundred years. There are many types that look similar but have slightly different dimensions. If you buy used equipment or have old sewing supplies hanging around then pay close attention that you have the correct bobbin for your machine and that some different type did not find its way into your bobbin drawer.

Home bobbin types

Class 15 - Class 15 bobbins are used in over 75% of all home sewing machines in use today. Class 15 bobbins come in two types as follows:

- **Class 15** - This is the original type with flat sides that is used in all machines with HA-1 style side-loading vertical bobbin-cases. These bobbins come in metal or plastic. The original Singer number is Class 15 and the same thing from an industrial supplier is called type A. If you buy quantities of prewound bobbins the sellers will call them type A bobbins. Do not use original flat sided class 15 bobbins in top loading machines. You may get away with it for a while, but you will eventually jam the machine.
- **Class 15J or SA156** - This type comes only in plastic. The Singer number is 15J and the Brother number is SA156, they are the same thing. This type has slightly curved sides (on the top and bottom) and slightly rounded edges. You can easily see the curvature if you hold the bobbin on a flat surface. This type is designed for machines with top-loading bobbin-cases. The slight curvature and rounding of the edges is so that as the thread passes over the bobbin (as it does in a

top loading machine) it does not catch on the bobbin. Prewound type A disposable bobbins seem to work well in most top loading machines, but you should test them in your machine first before buying a large quantity to make sure they work properly. You should also ask the supplier. Class 15J or SA156 bobbins can be used in older machines as a substitute for the original class 15 bobbins with no problems.

Class 66 - Class 66 bobbins are used in many Singer models made between 1930 and 1990, but are used in very few new models. Class 66 bobbins hold about 25% less thread then Class 15 bobbins. They come in both metal and plastic. All class 66 bobbins have curved sides.

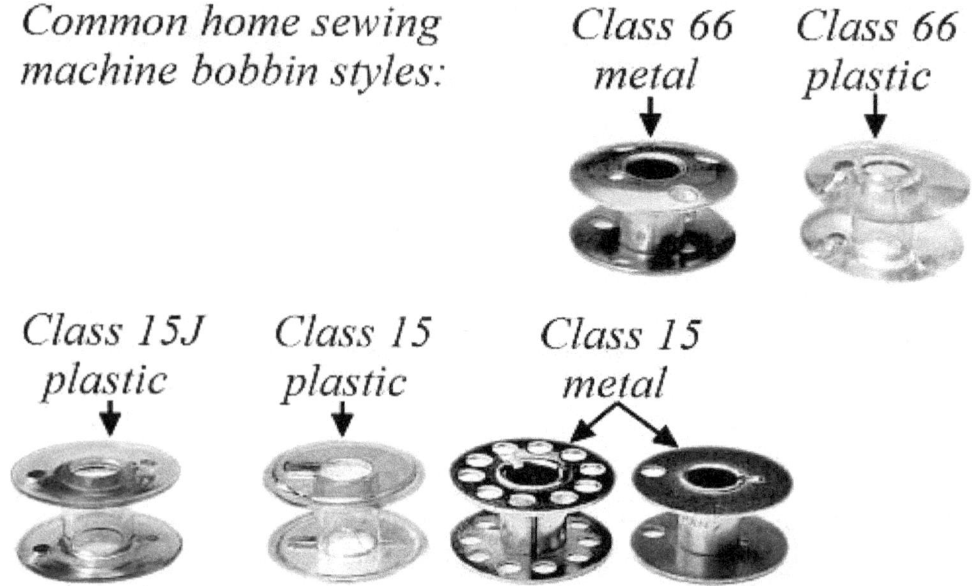

Other home bobbin types - Some of the premium brands like Pfaff and Bernina make machines that use brand specific (non-standard) bobbins that can only be used in their machines. Most home long arm quilting machines use the larger industrial "M" style bobbins. Some heavy-duty straight stitch home machines and some premium home machines use industrial "L" style bobbins.

Some home sewers use type L industrial prewound bobbins in home sewing machines that normally use Class 15 bobbins and report no problems, I have not personally tried this. The reason why they do this is that some prewound bobbin suppliers do not carry Class 15.

Industrial bobbin types

Most industrial bobbin types have letter names, A, L and M are popular types. Industrial bobbins are used in some high speed home straight stitch machines and in some quilting machines. Each type has different dimensions, use the correct type for your machine because they are not interchangeable. Watch for poor quality, it is a problem with industrial bobbins as well as home bobbins. The most common industrial bobbin types are L and M. The L bobbins are smaller. The M bobbins are much larger.

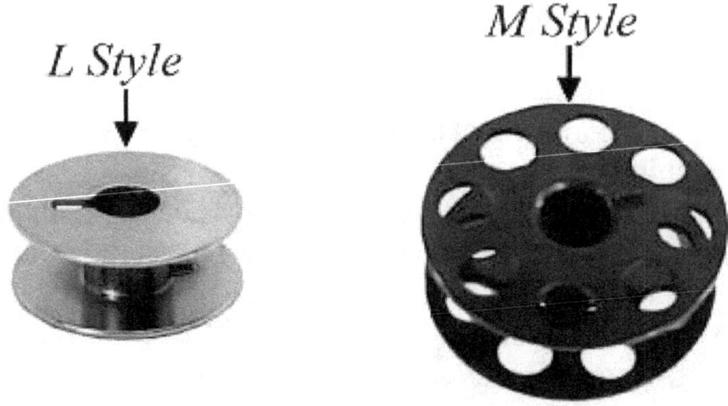

Testing a bobbin in the machine

This test will make sure that the bobbin can turn freely in the machine when the bobbin thread is pulled.

1. Make sure the bobbin has some thread but is not over filled. Make sure that the center end of the thread is cut off so there is no center tail of thread sticking out of the bobbin.
2. Insert the bobbin correctly into the machine and thread the machine so that the bobbin thread is coming out of the hole in the needle plate.
3. Grab the thread coming out of the hole in the needle plate and slowly pull out about a foot or two of thread.
   - Does the thread pull out smoothly but with some tension? If so then continue.
   - If you can see the bobbin on your machine, does it turn freely? If the bobbin turns freely and the thread pulls smoothly with constant tension then your bobbin is probably OK.
4. If the bobbin does not turn freely or the thread does not pull smoothly you may have a defective bobbin. Try another bobbin.
5. Make sure the bobbin is not over filled (see the section "Proper bobbin winding" later in this chapter).
6. Still not free and smooth? - See the chapter Troubleshooting. You must find and fix the problem or the machine will not function correctly. Usually changing to a good bobbin will fix the problem, but it could be something else.

How to inspect bobbins

The following is a checklist of bobbin inspection items and issues to watch for, after the checklist is a picture showing some of the common problems. Replace any bobbins you find that have issues or defects.

Existing thread – Never wind thread onto a bobbin that already has thread on it. If there is thread already on a bobbin and you need to use a different color or thread type then use a different bobbin or remove the existing thread from the bobbin before you wind new thread. The reason is that the end of the existing thread may get tangled in the new thread and jam your bobbin or cause other problems and also the new thread will not wind evenly over the old thread and therefore will not unwind evenly during use and will cause tension problems.

Center tail – After you wind a bobbin you must cut off the end of the thread that is coming out from the center of the bobbin. This must be cut off close to the bobbin so that there is no tail of thread sticking out. A thread tail can interfere with bobbin rotation and cause tension problems.

Over filled - Make sure your bobbins are not over filled with thread, see the section "proper bobbin winding" later in this chapter.

Warped – To test for warpage, put the bobbin on your machines bobbin winder and run the machine at a low speed. As you watch the bobbin spinning around, it should not tilt from side to side. The sides of the bobbin should be centered and not move orbitally. Most bobbins are warped a tiny bit, but when you see a warped one you will know it. Test a bunch of them at a one time so that you learn what a warped one looks like.

Sharp edges or imperfections - This will make the bobbin stick in the bobbin-case and cause erratic stitching and tension problems. Feel the bobbin with your fingers to find imperfections. Feel the bobbin as it is spinning on the bobbin winder when you are checking for warped bobbins. Do a visual inspection by looking at your bobbins closely.

Dimensions out of specification - Some bobbins are not manufactured correctly and have some part of them that is too big or too small. In all cases you can compare the bobbin you suspect to another bobbin or measure the bobbin with a micrometer.

- **Bobbin too tall** - This can make the bobbin stick or can prevent the thread from unwinding and cause tension problems.
- **Outer (flange) diameter too big or too small** - Too big will cause the bobbin to stick in the bobbin-case. Too small will cause the bobbin to rattle in the bobbin-case and make noise.
- **Hole diameter too small** - If the hole diameter is too small the bobbin will not mount on the bobbin winder.

Metal bobbins coming apart – Some metal bobbins are made in three sections and then crimped together like a can of tuna fish. Sometimes the crimps are not tight so the sections can rotate or move separately. To test for this just grab the top and bottom of the bobbin and twist in opposite directions. If the sections are loose you will be able to rotate the top or bottom. Nothing should move separately or rotate and the bobbin should feel like one solid unit.

Plastic bobbins cracking or warping - Plastic bobbins will occasionally crack, particularly if they are old because old plastic becomes brittle. Plastic bobbins can crack and split apart when they are wound very tightly. Look for cracks when you inspect your plastic bobbins before winding or using them. If you see cracks then discard the bobbin and get a new one.

Inspection - It is a good idea to periodically inspect (like before winding) all bobbins to make sure that they are OK.

Bobbin Problems:

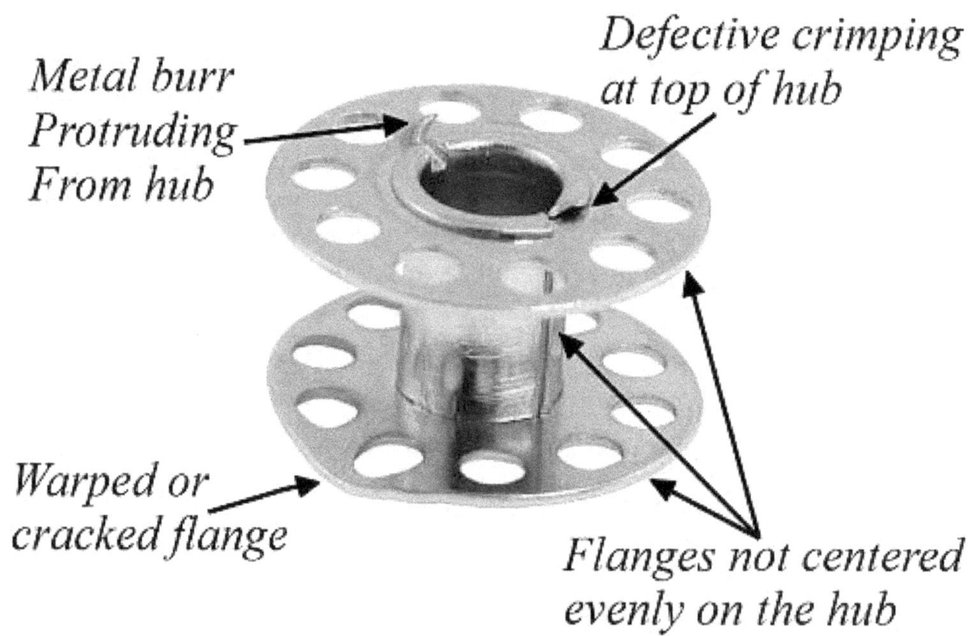

Plastic vs metal

Some machines use bobbins that come in either metal or plastic. Metal bobbins tend to be a little stronger and last longer, but there is not a big difference if both are of good quality. More important than plastic vs metal is getting a good quality bobbin. A good quality plastic bobbin is much better than a poor quality metal bobbin and vice versa.

Plastic bobbins can crack and split apart when they are wound very tightly. Metal bobbins can rust or become bent. Sometimes you can scrape the rust off with a small screwdriver and wipe any remaining rust off with a rag, if the rust is bad the bobbin should be discarded. If you spin the bobbin on the shaft of a small screw driver you can tell if it is bent. If you have a bent bobbin it should be discarded.

Proper bobbin winding

Incorrect bobbin winding causes a lot of the more nasty tension issues and it only takes a few minutes to learn to wind bobbins properly and avoid these frustrating tension problems.

Empty bobbin - Only wind thread onto an empty bobbin. Winding thread onto a bobbin that already has thread on it will cause uneven packing of the thread and can cause tension problems and jammed bobbin thread.

Bobbin winding tension - Bobbins should be wound with the correct amount of tension.

- Too much tension during winding can cause warping or breakage of the bobbin and will cause the thread to unwind erratically during sewing causing stitch problems.
- Too little tension during winding can cause the bobbin to have an unbalanced wind and shifting of sections of the wind (the thread on the bobbin). This can cause stitch problems while sewing.

Checking bobbin winding tension - Thread the machine as you normally would for bobbin winding. Pass the thread through the hole in the bobbin as you would to start winding, but do not run the machine, instead pull the thread and feel the winding tension on the thread. Now load another bobbin into the bobbin-case of the machine and draw the thread up through the hole in the needle plate (as you would for normal sewing). Pull the thread from the needle hole and compare that tension to the winding tension. The winding tension should be about the same as the bobbin tension of the machine.

Even winding - Watch the bobbin as it is winding make sure that the thread is evenly winding (usually from side to side). If the thread is not evenly winding then adjust the thread guides to fix the problem.

Correct filling - Do not over fill the bobbin. Over filling will cause severe tension issues or thread breakage. The thread on a wound bobbin should be at least 2mm below the rims of the bobbin at its highest point.

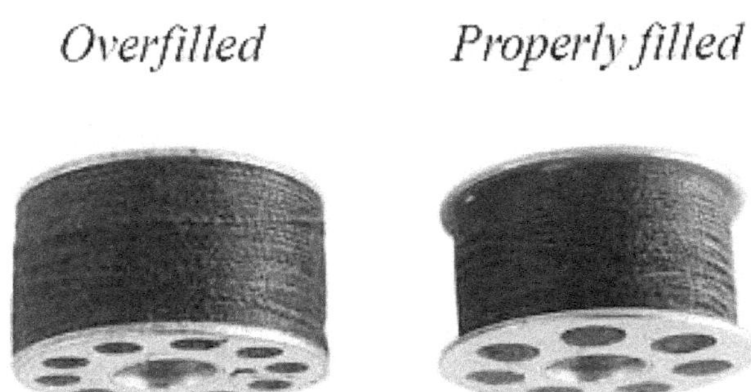

Presser Feet

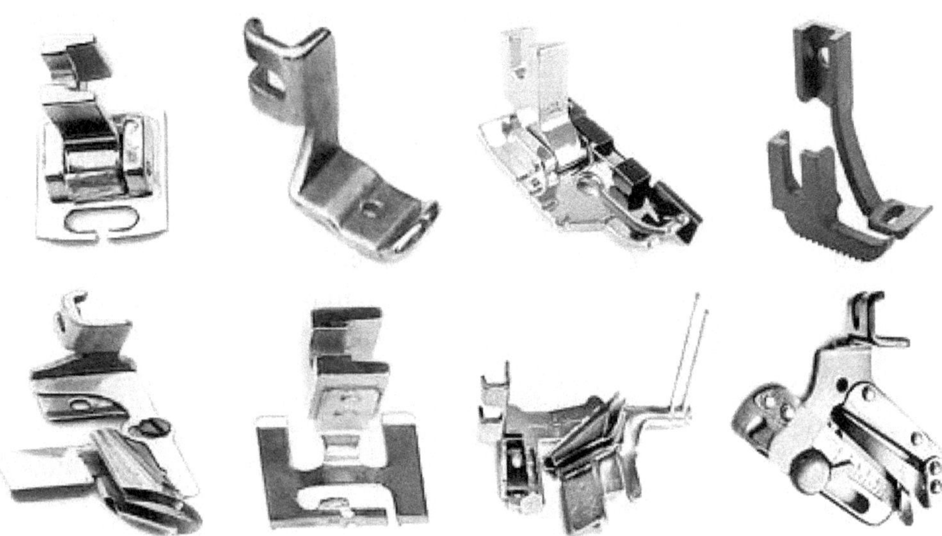

Presser foot basics

The presser foot is the part of the sewing machine that presses down on the fabric to hold the fabric in contact with the feed dogs. The feed dogs are mounted in the bed of the machine below the presser foot. The feed dogs have claws to grab the fabric, they move forwards or backwards to pull the fabric through the machine.

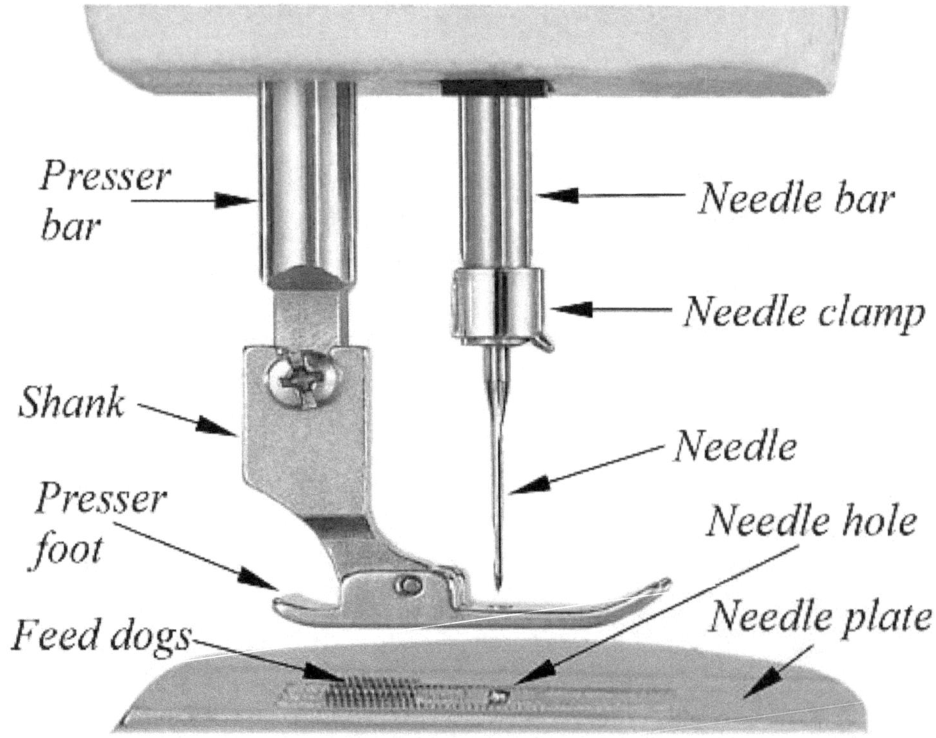

Most sewing machines come from the factory with a general purpose presser foot installed. The general purpose foot will work for most sewing tasks, but occasionally you will need to do something that a general purpose foot can not do, or can not do well. In that case you will need to use a foot made specifically for the task at hand. An example of specialized feet are button hole feet and piping feet. There are many kinds of specialized feet and accessories, we will go over the more common ones later in this chapter. Some examples of specialized feet are in the picture at the beginning of the chapter. More than any other part of the sewing machine a wide selection of presser feet allows the machine to do many tasks and types of sewing.

The shank

The foot bone connects to the ankle bone and the ankle bone connects to the leg bone,,, actually this is a fairly good analogy for the sewing machine shank and foot. The sewing machine foot attaches to the shank and the shank attaches to the presser bar of the machine. Some feet are permanently attached to the shank (like the feet in the picture above) while other feet use a snap-on style shank and are able to be removed from the shank. Most new machines come with a snap-on style shank. With a snap-on style shank different feet can be popped on and off easily without using tools. There are several different snap-on shank types from different manufacturers, but most manufactures are now using a universal snap-on shank.

The shank mounts to the presser bar of the sewing machine with a screw on most machines. On some machines this is a thumb screw that can be removed without a screwdriver and on other machines it is a machine screw and a screwdriver is needed. In most cases the machine screw can be replaced with a thumb screw if you prefer to be able to remove the shank without a screwdriver and have a machine that came with a machine screw.

Shank types for home machines

There are three standard shank types for home sewing machines; low shank, high shank and slant shank. These types are not interchangeable, you must use the correct type for your machine.

There are also some odd-ball shank types from manufactures that want to be different, usually to prevent you from using standard inexpensive feet and force you to buy feet from them at a high cost. Most of these manufactures will say that they do not use standard feet because they have a superior high technology design and that they are doing you a favor by selling you an $8 foot for $60.

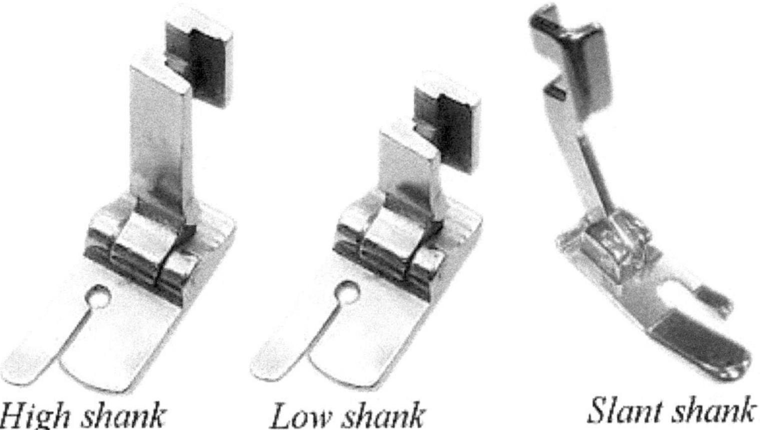

High shank *Low shank* *Slant shank*

Low shank - This is the most common type of foot and is used in more than 80% of all home sewing machines. A wide variety of low shank feet are available at inexpensive prices. There are also high quality straight stitch low shank feet available on the used market for very good prices such as ruffler, tucker and adjustable hemmer feet. These feet are made by Singer, Greist and others. They can be used on both straight stitch and zigzag machines (as long as the zigzag machine is in straight stitch mode with the needle set to the center position.

High shank - This type of shank is used in some home machines. A high shank foot has a longer shank and the mounting screw to attach the shank to the presser bar is higher up (this is the reason it is called high shank). This higher mounting screw gives more clearance around the foot for attachments and the use of rigid templates or embroidery hoops. There are “high shank to low shank” adapters to extend the presser bar and allow the use of low shank feet with a high shank machine. These can be found on eBay and through other on like sellers. There is a large selection of high shank feet made for industrial machines. These feet will also work on high shank home machines and are priced very inexpensively. There are some industrial sellers that offer large assortments of feet at extremely good prices on eBay and through other on-line sites.

Slant shank - Singer made slant shank sewing machines for many years and these machines use a slanted shank. It is called slant shank because the needle bar and the presser bar of these machines are slanted toward the front of the machine. This was done to move the needle and presser foot further forward for better visibility. Even though these machines are no longer made, there is a large selection of slant shank feet available on the used market.

Snap-on home feet

Many new zigzag machines come with a universal snap-on style shank and can use universal snap-on feet. These include most machines from Brother, Singer, Janome and many other brands. Any low shank or high shank machine can use a universal snap-on shank to replace the existing shank so that you can use universal snap-on feet. If you have a low shank machine make sure you get a low shank version of the snap-on shank (or high shank if you have a high shank machine).

For slant shank machines, Singer did make a snap-on slant shank and assorted snap-on feet, but they are not compatible with the universal snap-on feet that work with high and low shank machines, the feet have a different coupling system and can not be interchanged.

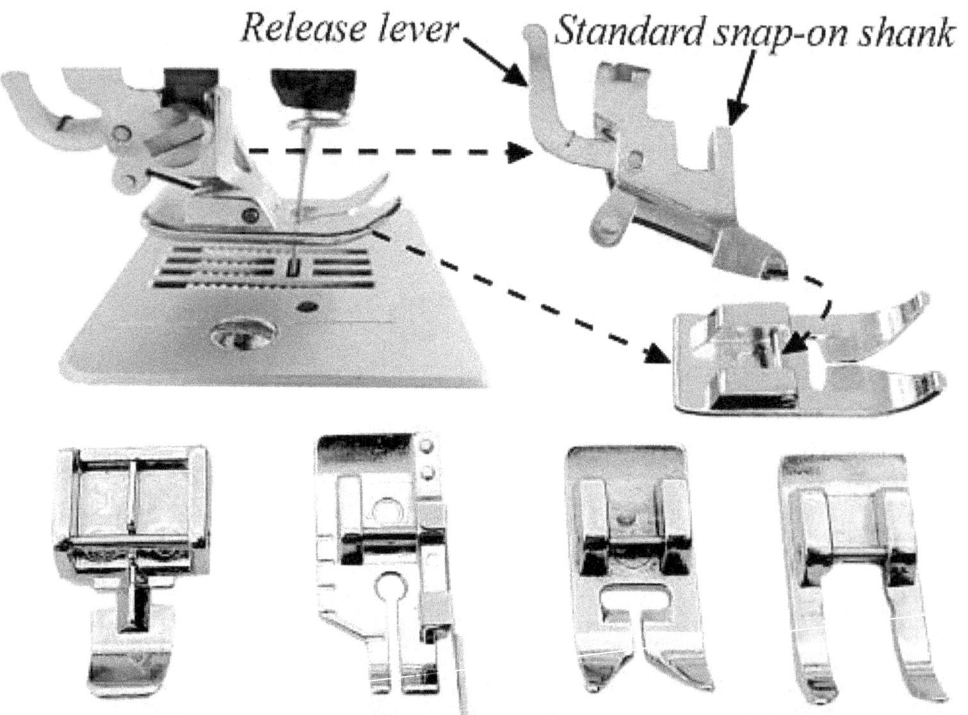

Many types of snap-on feet are available

Foot considerations for zigzag machines

Presser feet come in zigzag and straight stitch types. Zigzag feet have a needle opening that is elongated from side to side to allow the needle to move from side to side during zigzag stitching.

Using older zigzag feet on newer machines - Make sure that the needle opening is wide enough for your machine. Old zigzag machines have a narrow zigzag width and feet for these machines have a smaller needle opening then the feet for newer machines. If you use these older feet on a newer machine the needle may strike the foot. To check if you have proper clearance just set the machine to the widest zigzag stitch and slowly rotate the hand wheel while looking at the needle to see that the needle does not touch the foot and has adequate clearance (at least 1mm of clearance to the sides).

Using straight stitch and specialized feet - Straight stitch feet and most types of specialized feet have a needle opening that is round or not elongated from side to side. These feet can not be used for zigzag stitching because the needle will collide with the foot and break. If you use these feet with a zigzag machine, the machine must be set to the straight stitch mode and the needle must be in the correct position so that it comes down in the opening of the foot. Some zigzag machines have a needle position control and you can select from left, center or right needle positions when the machine is set to the straight stitch mode (or zero stitch width).

Checking clearance - It is a good idea when attaching a straight stitch foot or any specialized foot to a zigzag machine to manually test the needle clearance before sewing. This is done by slowly rotating the handwheel of the machine by hand and observing the needle as it lowers and raises to make sure that there is adequate clearance between the needle and the foot.

Considerations for needle position - Some zigzag machines default to the left hand needle position when in straight stitch mode and must be manually set to center position when using a straight stitch foot. Some electronic machines will revert back to left hand needle position every time the machine is turned off and turned back on again, if you have one of these machines don't forget to reset the needle position every time you turn on the machine or the needle will collide with the foot and break when you start to sew.

Foot specifications

There are several specifications other than the shank type that should be correct for the foot to function properly. If you use feet that are designed for the specific model of machine that you are using then these specifications will be correct, but if you are using a general purpose or generic foot you should pay attention that the following foot specifications will work with your machine.

- The foot should be wide enough cover the feed dogs.
- The foot should be long enough to cover the feed dogs or at least most of the feed dogs.
- The needle opening in the foot should be centered on the needle and the needle should never contact the foot. There should be more than .5mm of clearance between the needle and the foot even when the foot is tilted to the extreme forward and backward position.
- For zigzag machines the needle opening in the foot should provide clearance for the needle for the full range of side to side needle movement during zigzag stitching at the maximum width setting.

Foot types

The following is a list of the most commonly used types of feet. The feet are pictured in close up views with the foot not mounted on a sewing machine to show the details of what the feet look like, so you can clearly see the whole foot and see some of the features. I did not show the feet in use on a machine with fabric, because the fabric would obstruct the view. There are many sewing books that go through the operations such as gathering or piping and have pictures of the feet in use with fabric, one great source for this is the instruction manual for the Singer 15-88 sewing machine that can be downloaded for free from Singer at http://www.singerco.com/accessories/instruction-manuals just put in the model number (15-88 in this case) and hit the search button then download the PDF file. The section on feet starts on page 36 and shows the feet in use. Actually Singer has put most of its manuals on the site for free download and some of the newer manuals for the 500, 600 and 700 series machines have good sections on feet as well.

General purpose zigzag foot - Used as the main foot on zigzag sewing machines for all basic utility sewing. This foot can be used with all stitch types.

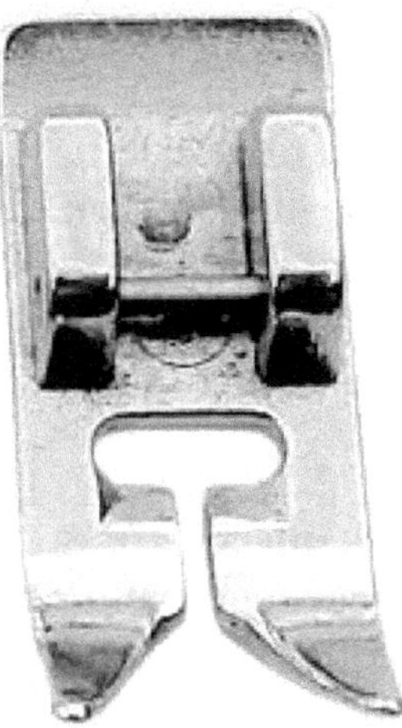

The picture above shows a snap-on version of a zigzag foot.

General purpose zigzag with locking pivot - Same as the general purpose zigzag foot except for the addition of the locking pivot. When the pivot is locked the foot will remain parallel to the needle plate at all times. The pivot is used to lock the foot for going over thick seams. This is handy for sewing jeans, jackets, tents, backpacks, cases, etc.

General purpose straight stitch foot - Used as the main foot on straight stitch sewing machines for all basic utility sewing. This foot may also be used on zigzag machines, but only when sewing in straight stitch mode (never in zigzag mode or needle breakage will result).

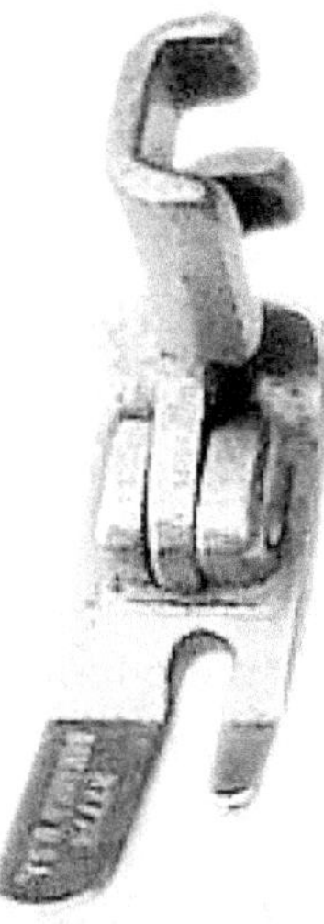

The reason to use a straight stitch foot on a zigzag machine is for better visibility, better cornering and for clearance when sewing in tight places. In the picture above a low shank version of a straight stitch foot is shown.

Adjustable hemmer - Used for hemming the edge of fabric, the adjustable hemmer can make medium to wide hems.

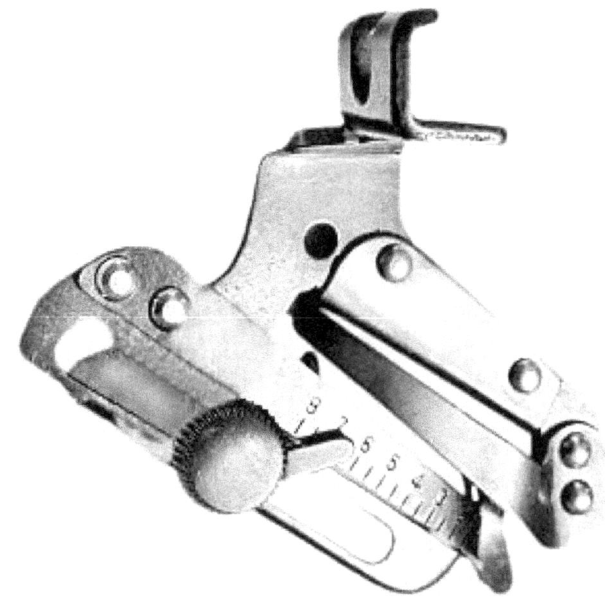

An adjustable hemmer made by Singer is shown above.

Binder - Used to apply folded and unfolded bindings and bias tape to seams and the edge of fabric.

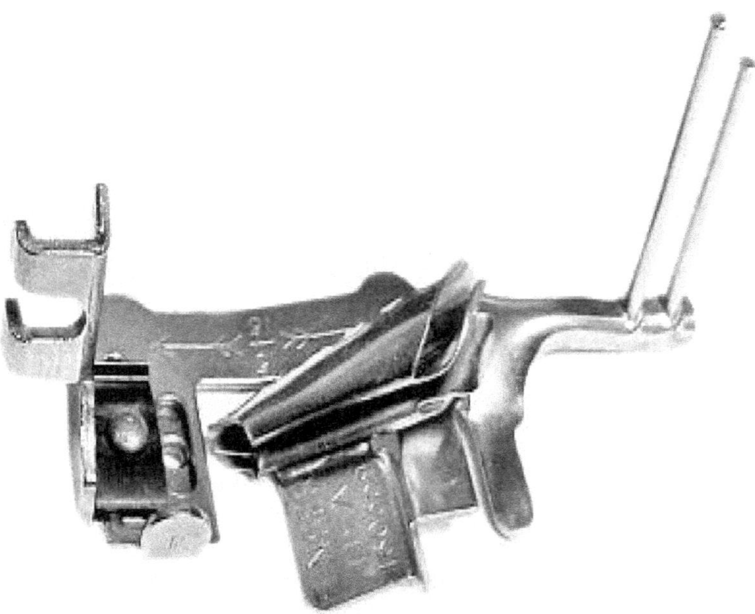

Some binder feet can handle only certain sizes of bindings, other types have multiple slots or are adjustable for a variety of sizes. The binder foot shown below has adjustable stitch position from side to side. It also features multiple slots for various size bias tape, but the slots are on the other side of the former so you can not see them in the picture.

Button hole - The button hole foot has a sliding action and works best with sewing machines that have a built in buttonholer function.

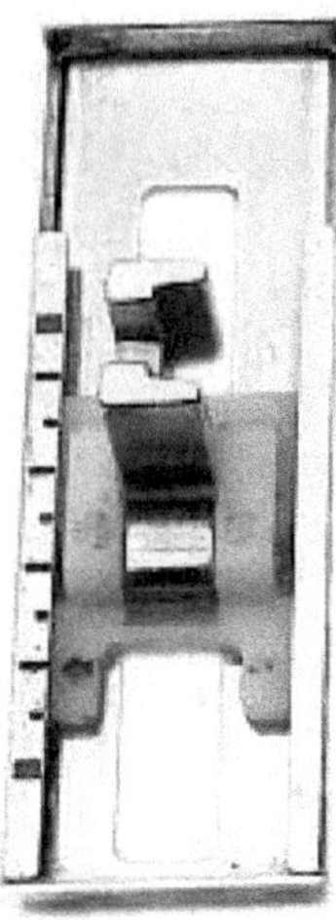

The button hole foot can also be used for bar tacking.

Compensating - Stepped design provides consistent feeding when sewing parallel to seams with layers that are of different heights.

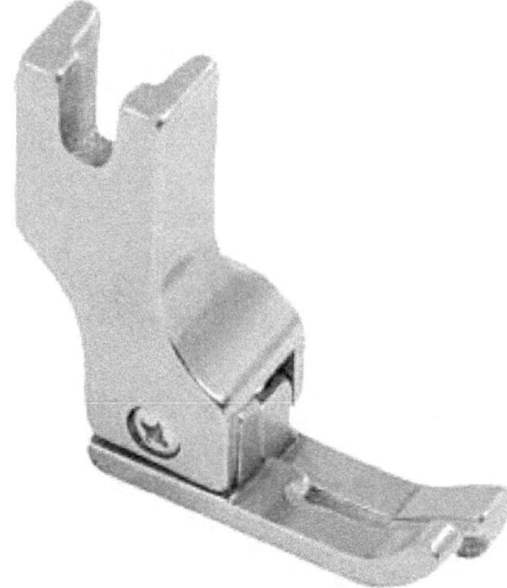

If you look in the picture above you can see that the left side of the compensating foot is higher then the right side of the foot. Some compensating feet are adjustable and can be adjusted so that either side can be raised or lowered.

Cording and piping - Used for attaching cording and piping.

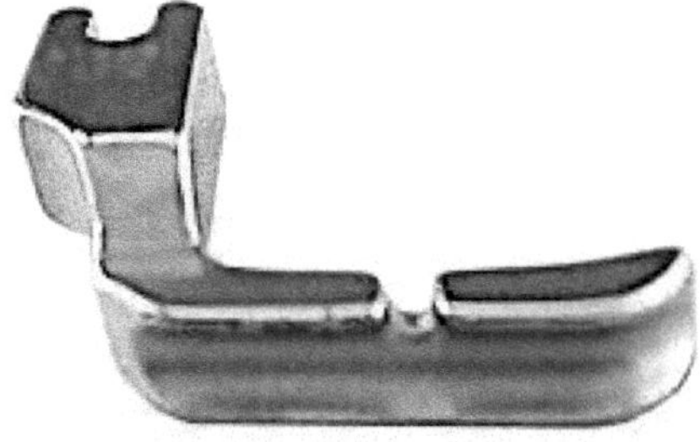

This foot has a channel on the bottom of the foot running from front to back for the the cording or piping to go through. This channel helps center the needle perfectly on the edge if the piping as it goes through the foot. Piping feet come in various sizes or you can buy sets of them with a range of sizes.

Darning foot - Used for darning and also can be used for free motion quilting.

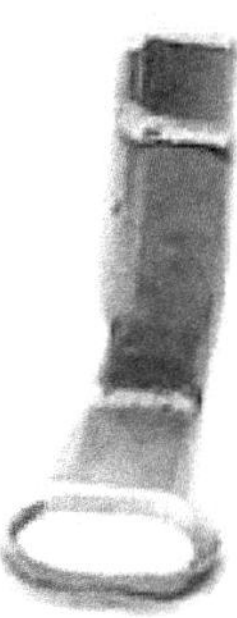

The darning foot provides better visibility than a general purpose foot. Darning feet are made to be used with the feed dogs disengaged (The feed dogs can be lowered on some machines or a feed dog cover plate can be used). Since the darning foot is only used to hold the fabric down during stitching and does not hold the fabric into contact with the feed dogs, a much lower presser foot pressure can be used to allow easy movement of the fabric through the machine (on machines that have adjustable presser foot pressure).

Edge cutting foot - Also called a side cutting foot, this foot has a built in cutter that is powered by the needle bar of the sewing machine.

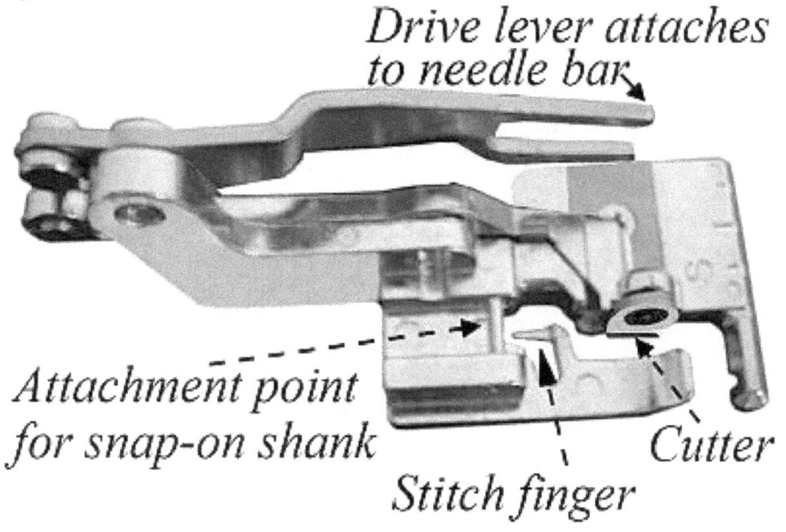

The edge cutting foot is used for edge finishing and cutting in one operation. The model shown above is a Brother SA177 and also has a stitch finger that is similar to a serger to create a more consistent overedge stitch for edge finishing fabric.

Edge stitcher - The edge stitcher is used to stitch accurately at a precise distance from the edge of the fabric or very close to the edge of the fabric.

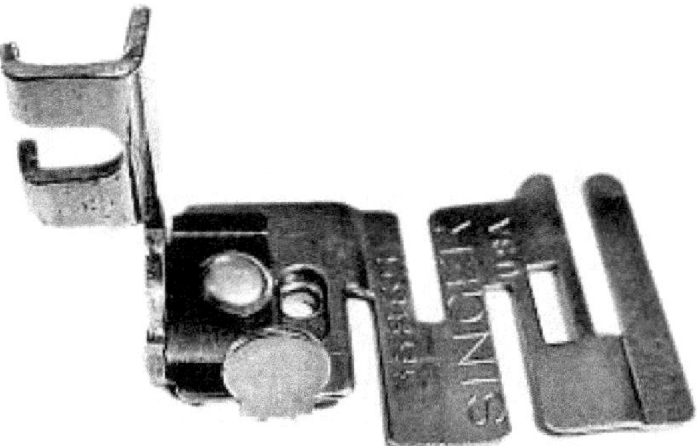

The edge stitcher has open end slots that are spaced at various distances from the stitch line and you insert the fabric through the slot that corresponds to the spacing you need. The rectangular closed slot is for feeding tape while edging.

Embroidery foot - Similar to the darning foot, but has a built in pressure spring system to apply a consistent low pressure so as not to cause the fabric to stretch.

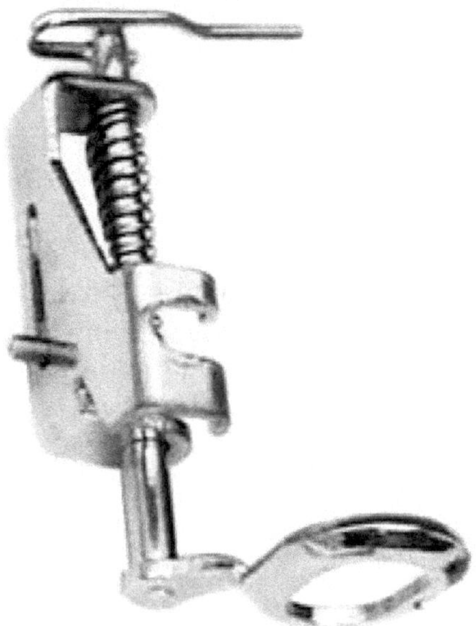

The embroidery foot must be used with the feed dogs disengaged or with a feed dog cover plate.

Gathering foot - Used for shirring and gathering.

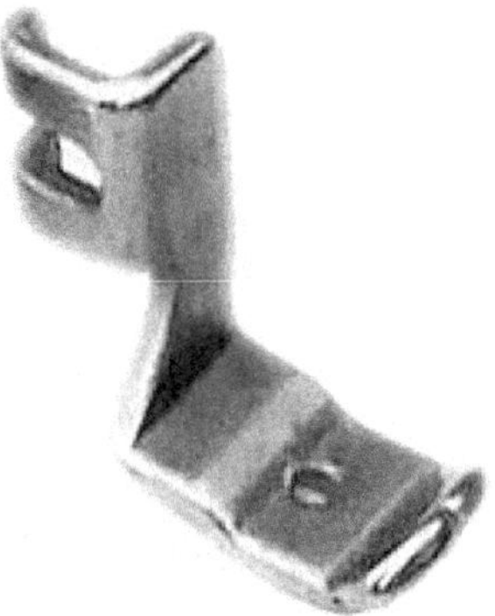

The amount of the shirring or gathering is controlled by the stitch length of the machine, a longer stitch length results fuller shirring and more gathering. A Singer version is shown above, models from other companies may look quite a bit different, but perform the same function.

Hemmer foot - Used for hemming the edge of fabric.

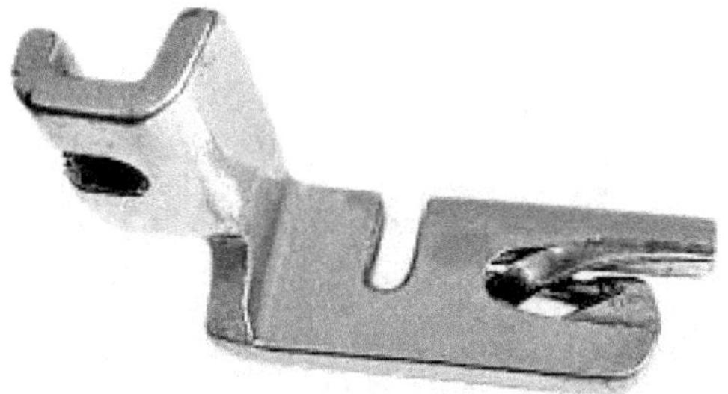

The hemming foot shown above can make narrow hems and felled seams.

Piecing - Used for precisely joining sections of material.

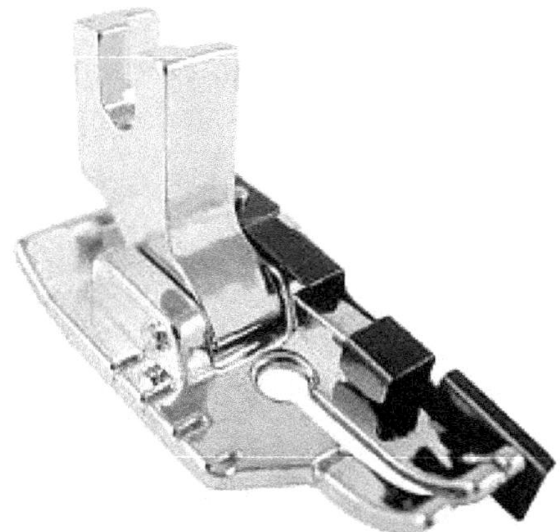

The piecing foot places the seam a set distance from the edge of the fabric. Piecing feet are available in 1/8Th inch and 1/4Th inch widths. A 1/4 inch version is shown above.

Pipping - Used for piping. Same as the cording foot above.

Quilting - Used for straight stitching quilted or fluffy materials, the quilting foot is wide like a show shoe so it will not sink in and stick with fluffy materials.

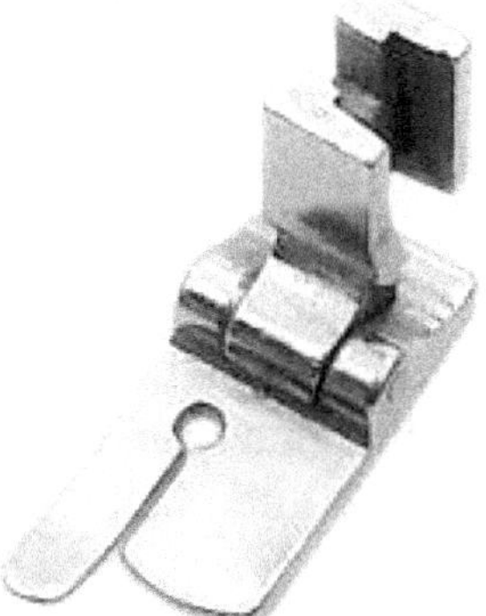

There are other feet that look different but have a similar function and are also marketed as quilting feet, so don't be surprised if you see some strange looking quilting feet! Notice that the version in the picture above has a left side needle hole so it can only be used with zigzag machines that have a needle position control or a left side needle position. There is another version of this foot with a center needle position for straight stitch machines and zigzag machines with a center only needle position.

Roller - The roller foot has rollers that help insure consistent feeding of materials with sticky surfaces such as vinyl, plastic and some leathers.

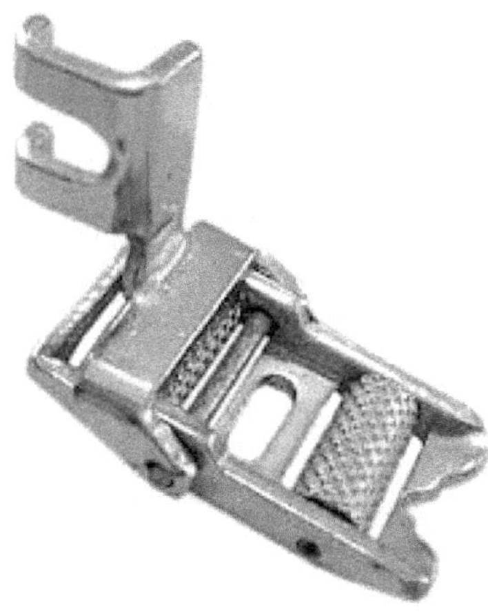

The roller foot pictured above is an older all metal version, most newer snap-in style roller feet have a plastic body.

Ruffler foot - The ruffler foot is quite flexible and can be used to ruffle, gather, plait, attach facings, ruffle and pipe in one operation, ruffle and attach facings in one operation or for group plaiting.

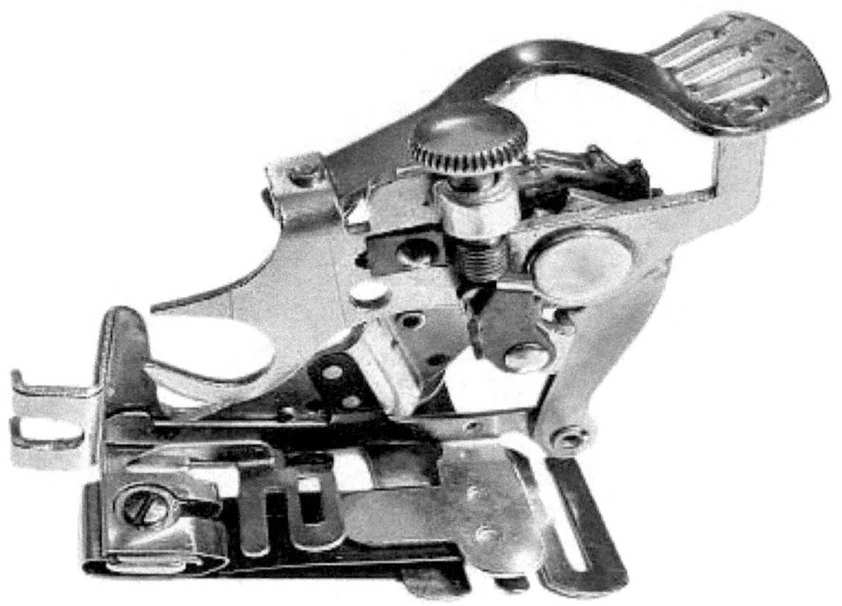

The ruffler foot looks complicated, but once it is set up properly very consistent results can be achieved. The mechanism is powered by the needle bar. If you buy a used one and are not familiar with its operation, make sure you can get an instruction guide for that specific model. It will be easy to set up and use with the instructions, but could be difficult to figure out without the instructions. The instruction manual for the Singer 15-88 sewing machine that can be downloaded for free from Singer has an excellent set of instructions for the ruffler foot pictured below.

Teflon foot - Similar to a general purpose foot, but made of non-stick Teflon instead of metal.

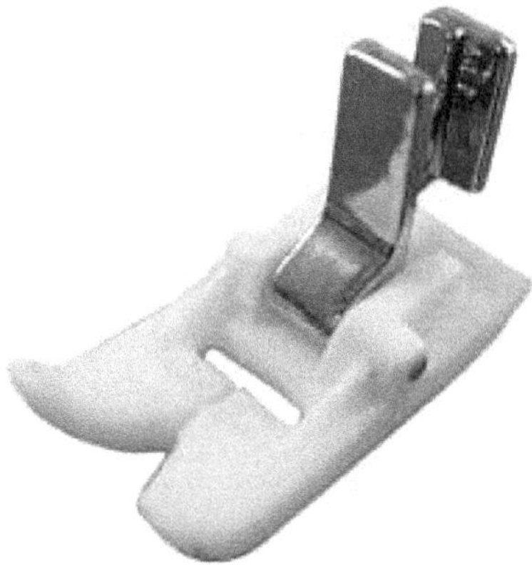

The Teflon foot provides better feeding with some hard to feed fabrics and materials. Can help reduce friction and provide better feeding of multiple layers of fabric.

Tucker foot - Use for tucking. The mechanism of the tucker foot is powered by the needle bar of the machine.

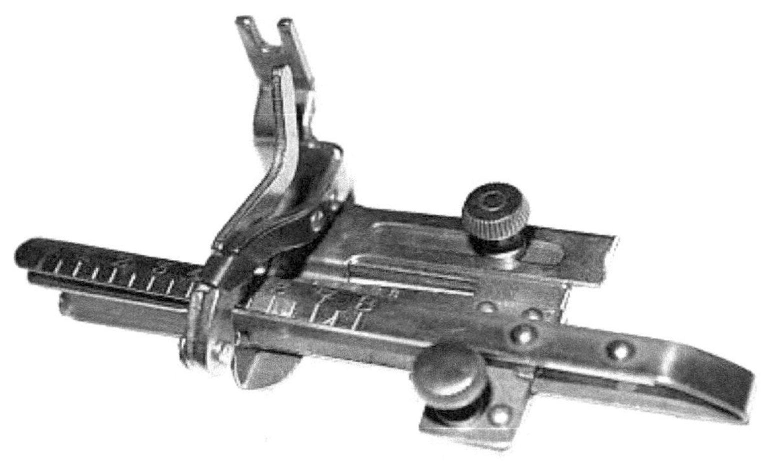

Zipper foot - Used to attach zippers.

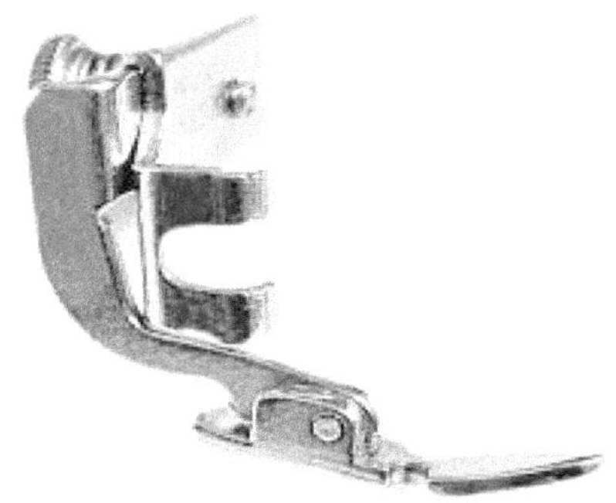

The zipper foot can also be used for pipping and close stitching in tight areas. The adjustment thumbscrew is used to change the position of the foot from side to side, the foot can be placed on either side of the needle.

Button Hole attachments

For machines that do not have a built in button hole function, a button hole attachment mounts to the shank of the machine in place of the foot and allows the machine to make button holes. Most button hole attachments come with several templates for making button holes of various sizes and shapes.

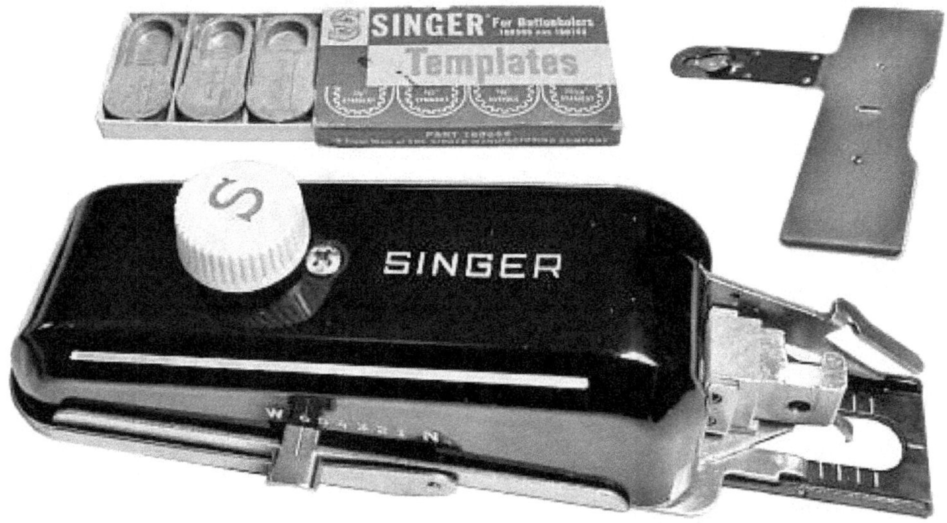

Singer made several models of high quality button hole attachments for machines with low shank feet, you can get these attachments on the used market (eBay) quite inexpensively (under $20).

YS Star (Yoshikoshi Mfg) is a Japanese company that makes new button hole attachments.

- For low shank home sewing machines the model YS-4454 has an adjustable length setting and has a removable screw for making bar tacks.
- For machines with standard high shank feet the model YS-4455 is heavy-duty and has several adjustments to control the width and length of the stitching. It also has a removable screw that causes the attachment to make bar tacks. The YS-4455 is pictured below.

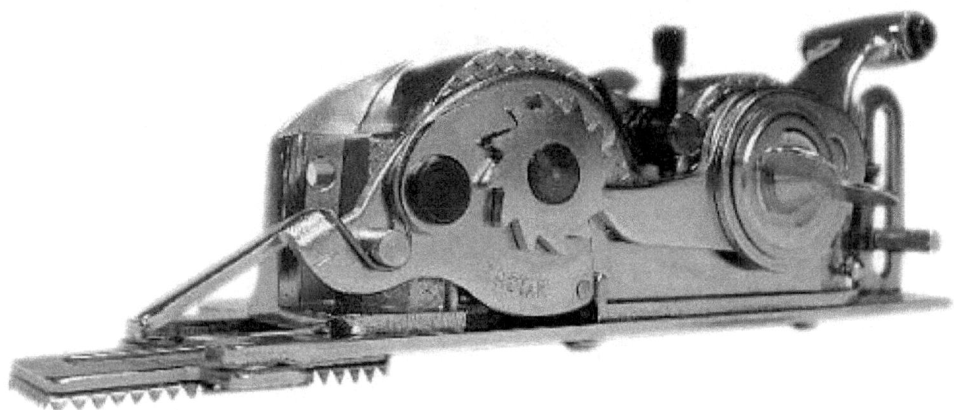

Walking foot attachments

Also called an "Even feed Foot" or "Plaid Matcher", these attachments help to feed difficult fabrics evenly through the machine. The name "Plaid Matcher" is used because they are often used to help feed plaid fabric so that pattern will match on seams. The Even Feed Foot is not a true walking foot like you find in an industrial machine. The attachment has two feet controlled by a mechanism that is powered by the needle bar of the machine.

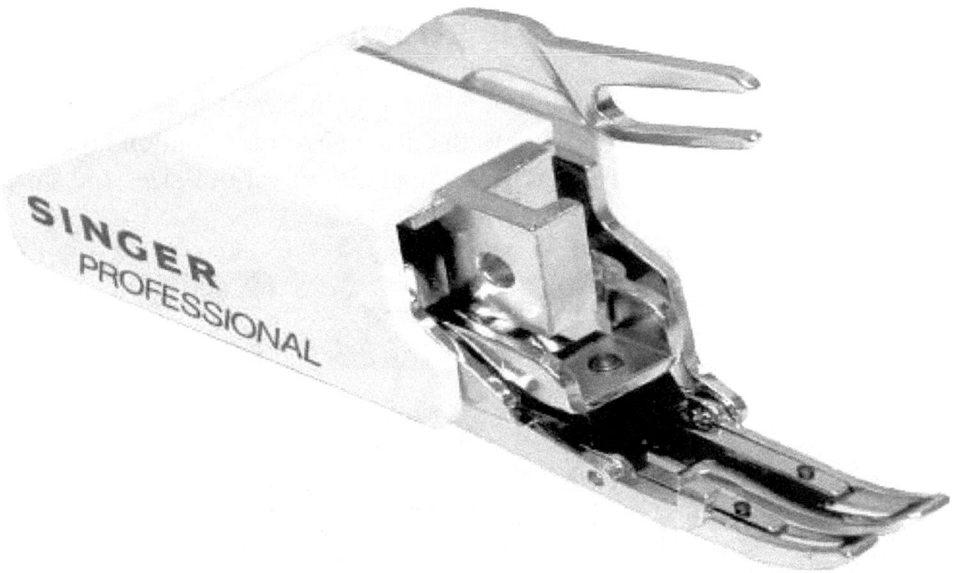

The plaid matcher foot works as follows:

- The first foot is lowered down into contact with the fabric and holds the fabric steady while the needle is penetrating the fabric.
- The second foot can move freely forward and back ward on bearings. This foot is lowered while the feed dogs are raising and allows the feed dogs to move the fabric forward with little resistance or friction.

- With a regular foot the feed dogs force the fabric to slide against the stationary presser foot to move the fabric forward. The feed dogs have to fight the friction of the fabric dragging against the bottom of the presser foot.
- With the plaid matcher the feed dogs move the fabric forward along with the second sliding foot that moves freely with the fabric. The fabric does not slide against anything and there is no friction between the fabric and the moving second foot.
- Because of this lack of friction, the feeding action is much more consistent and difficult to feed fabrics are much easier to sew.
- However, the attachment is not the same as an industrial walking foot and will not allow you to sew heavy leather, webbing or multiple layers of heavy fabric like you can with an industrial walking foot machine.

Zigzag attachments

The zigzag attachment is for use with straight stitch machines to enable zigzag and decorative stitch capability. In the model shown below pattern cams are used to select the from various stitches and decorative patterns.

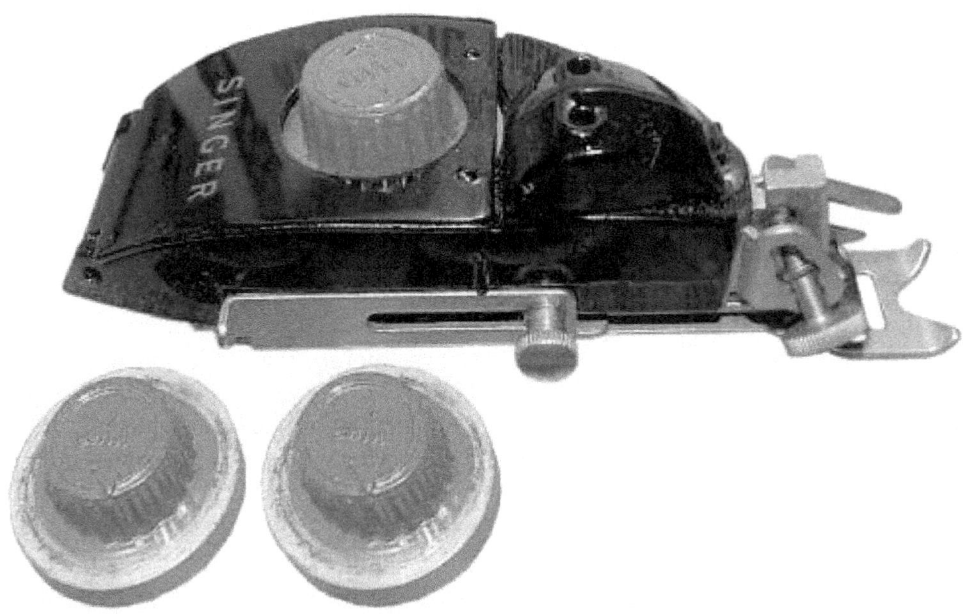

Fabric

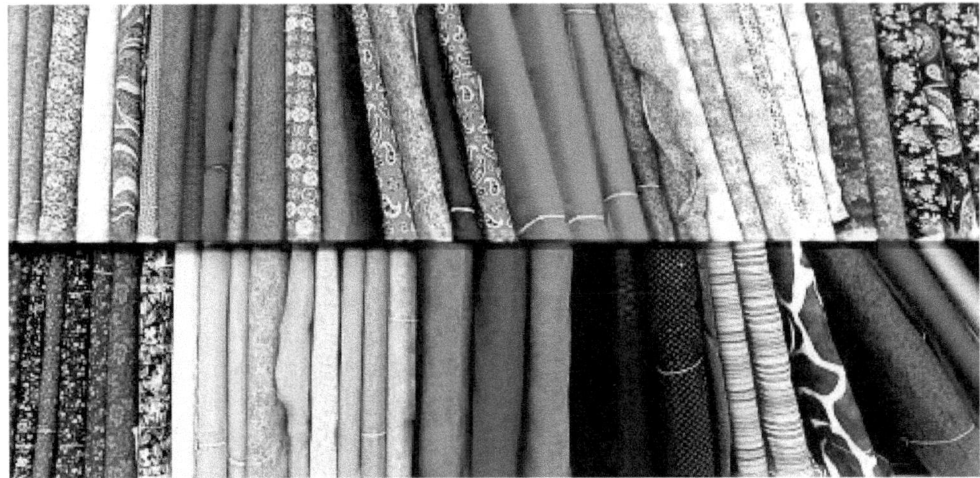

There is a wealth of good information available about fabrics on Wikipedia and the Internet, so I will just go over some of the important things in a condensed format.

Fabric weight

Fabric weight is the measurement of how much a certain quantity of fabric weighs such as one square yard. Since a yard is 3 feet, if you cut a square piece of fabric 3 feet by 3 feet that will be one square yard (a square yard is also called yd2, yd is for yard and the 2 is because it is squared). Likewise, if you cut a piece of fabric one meter by one meter, it will be one square meter (also called m2).

Units of measurement - Fabric weight is normally expressed in ounces per square yard which is written as Oz/yd2 or as grams per square meter which is written as g/m2. Grams per square meter is also written as GSM (Grams per Square Meter).

Warning! - There are other measurements you have to watch out for such as ounces per linear yard (Oz/yd) or grams per linear meter (g/m). These measurements are kind of stupid because they don't actually tell you the weight of the fabric with out knowing more information. A linear yard or meter is measured by rolling fabric from a roll or bolt and measuring how much fabric has been unrolled from the roll or bolt. If you unroll one yard, that is one linear yard. The problem with these measurements is that we don't know exactly how wide the fabric is! Fabric comes on rolls and bolts of many widths. You have to ask how wide the roll or bolt is and then you can do a calculation to convert from Oz/yd or g/m to Oz/yd2 or g/m2. Many people have received fabric that they thought was going to be a thick 16 Oz fabric (if it was measured in Oz/yd2) only to find they receive a thin 12 Oz fabric (that was listed as 16 Oz but measured using Oz/yd).

Buying retail - Buying fabric at retail stores can be frustrating because sometimes the fabrics are only marked with 6 Oz or 8 Oz or whatever, but they don't specify if it is Oz/yd2 or Oz/yd, so you really don't know what 8 Oz means, its basically meaningless! Sometimes the fabric is just labeled as medium or heavy with no weight at all. In these cases if I am really concerned and need to know the exact weight, I will buy a yard of fabric and then weigh it on a digital postal scale to find out the true weight.

Fabric types

There are many types of fabric available, generally they fall into three main categories - weaves, knits and non-woven fabrics. Listed here are the most popular types with usage recommendations.

Weaves

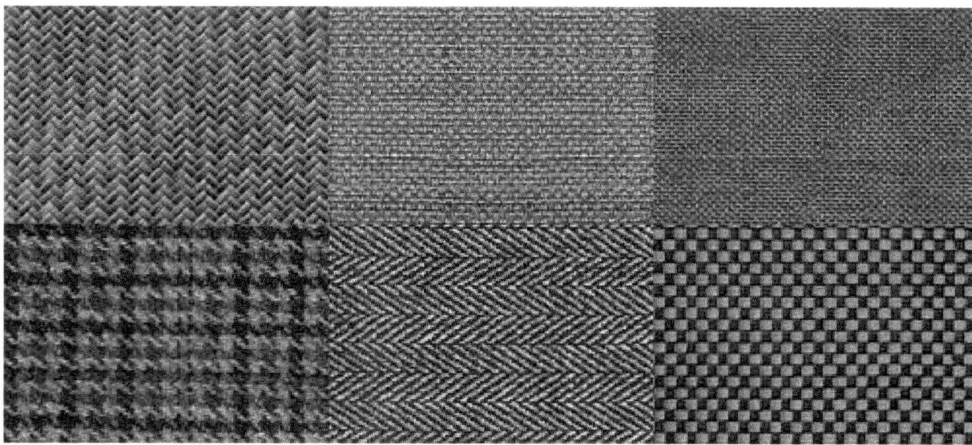

Woven fabrics are made from interlaced parallel rows of yarn or thread. The lengthwise threads are called warp threads and run for the entire length of the roll of fabric. The crosswise threads are called weft threads and run across the roll of fabric. The main types of weaves are as follows:

Plain weave - In the plain weave the warp and weft threads are intertwined symmetrically in a crisscross pattern of under-over-under-over. In a balanced weave the same thread size and type is used for both the warp and the weft. A balanced weave will stretch the same amount from side to side as it will lengthwise. If you look at a plain weave under magnification it will have a checkerboard pattern.

- Plain weaves are good for sheets & pillow cases, draperies, quilts, table cloths, upholstery.
- Plain weaves have high durability but do not stretch well, so they are not the best choice for clothing.
- Because plain weaved fabric is relativity flat, it is the best fabric for printing and comes in large variety of prints.
- Examples of plain weaves are poplin, taffeta, duck, canvas, chiffon, georgette, calico, percale, cheesecloth, etc.
- The basket weave is a variation of the plain weave in which two or three threads are used for the warp or the weft. The basket weave has an exaggerated pattern and is often used for wall papers and decorative uses. Types of basket weaves are oxford and monks cloth, etc.

Satin - In a satin weave there are a number of interlacing points that are skipped by the warp threads causing the warp threads to form a smooth and shiny surface. Satin is used for woman's clothing, sports shorts, some types of men's shirts, nightgowns, lingerie, men's ties, fancy bed sheets, draperies and couture. Examples of satin are crepe backed, brocade satin, duchess satin, chameuse, etc.

Twill - In the twill weave the under-over weave pattern is modified to skip over alternate warp and weft interlacing points, causing a diagonal ribbed pattern to be made. Twills contour well around curved surfaces making them popular for pants, jackets and upholstery. Some examples of twills are denim, drill, gabardine, tweed, chino, serge, etc.

Complex - With the invention of computer controlled weaving equipment, complex weaves became possible. Complex weaves can be based on any of the main weave types. Variations of the weave are introduced by the computer. These variations can be consistent or can be programmed to reproduce designs and images.

Knits

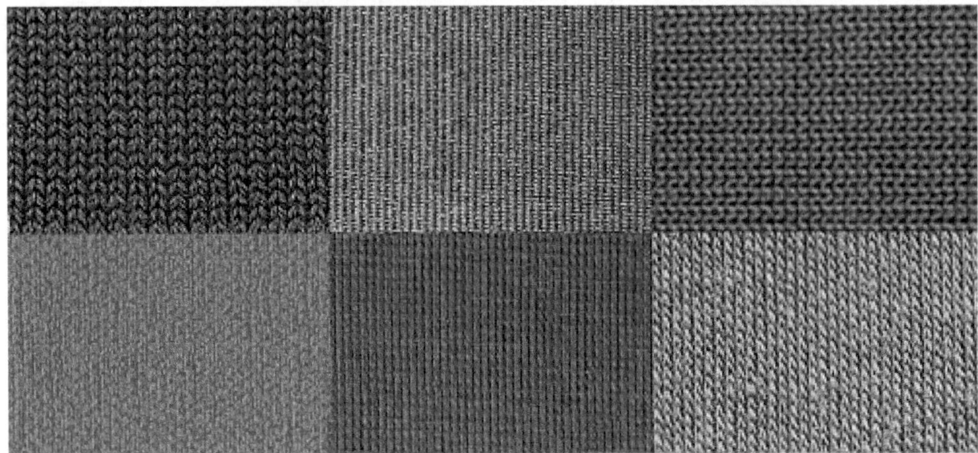

Knit fabrics are made by looping or braiding yarns instead of parallel interlacing as in a weave. Because of being looped the yarns are always curved and this causes a knit to be thicker, softer and more flexible then a similar weight weave. Knits can stretch in both directions whereas most weaves can stretch a smaller amount in one direction, if they can stretch much at all.

Weft knits - are made with weft yarns only (no warp yarn is used). The yarn is looped in successive rows with each row linked to the previous row. Examples of weft knits are jersey, purl, etc. Weft knits can be made by hand or by machine.

Warp knits - are made with many parallel yarns that are looped the vertical rows on either side. Warp knits are usually made by machine only. Some examples of warp knits are tricot, cardigan, milano, milanese, etc.

Double knits - look the same on both sides and do not stretch much. Double knits and are popular for sweaters, shirts, pants, jackets, dresses, sleepwear and socks. Most double knits are medium and heavy weight fabrics.

Single knits and jersey knits - look different on each side. Single knits and jersey knits are used for T-shirts, dresses, shirts, pants and sleepwear. Most single knits are light to medium weight fabrics.

Super stretch knits - are made with spandex are used for swim-wear, active wear, leotards, socks and leggings.

Sewing recommendations for knits - Knits can be difficult to sew with a sewing machine because they stretch a lot and are prone to feeding problems which cause erratic stitching. Here are some tips for sewing knits:

- Use a plaid matcher when sewing knits - A plaid matcher is an attachment that will help feed materials more consistently. Some home sewers and home sewing machine stores call a plaid matcher a walking foot, but it is really not a walking foot (see the chapter "Presser Feet").
- Use a serger to sew knits - When possible use a serger, most sergers have differential feed mechanisms that are designed to help feed knits and stretch fabrics more evenly.
- Use stabilizer - If you are not using a plaid matcher or a serger (or even if you are and have problem fabrics) you can use a stabilizer to help with even feeding of knits and stretch fabrics. A stabilizer is a paper like material that you use over (or under) the fabric. The stabilizer stops the fabric from stretching. Stabilizer paper can be purchased from fabric stores or you can use some

other thin paper like tissue paper, news paper or phone book paper. The stabilizer paper that you buy works better because it is easier to remove after the seam is made and leaves less bits of paper stuck in the seam that have to be picked out with tweezers after sewing. Actually if you can use a shorter stitch length, sometimes the closely spaced needle holes will cut the paper stabilizer allowing clean removal.

- Use ball point needles - Ball point needles give the best results with knit fabrics because they tend to spread the yarns and not cut through the center of yarns like a sharp tip needle does.

Technical fabrics

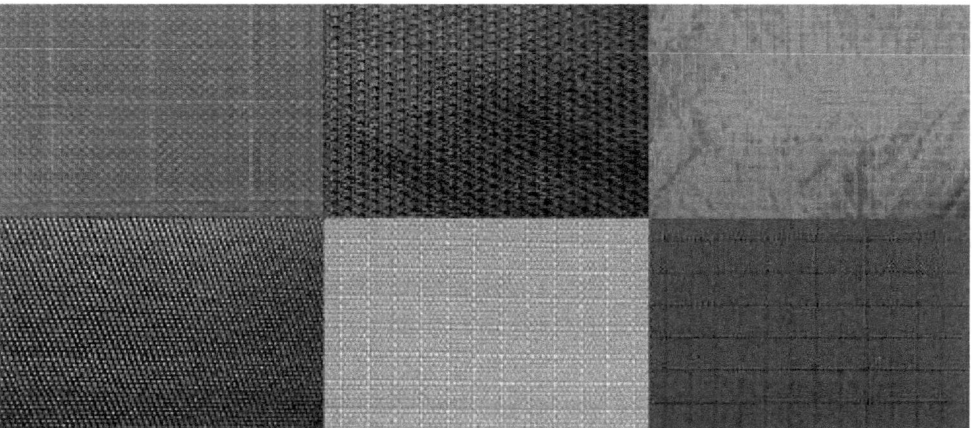

Technical fabrics are high tech fabrics and fabrics that excel at special applications such as extreme cold weather gear, high durability, rain protection, low friction sports wear, upholstery, structural elements, composites, etc. Some technical fabrics such as microfiber polyester have become popular for everyday clothing.

Rip-stop nylon and polyester - Several types of rip-stop nylon and polyester are shown in the picture above. Rip-stop fabrics are used in tents, rain gear, backpacks, luggage, camera cases, awnings and many other types of outdoor gear.

Polar fleece - Made from synthetic yarn derived from ultra fine polypropylene or polyethylene fibers. The yarn is knitted and then brushed. The brushing loosens up some of the fibers creating the frizzy (fuzzy) surface. Polar fleece retains very little water when wet and has very good insulating properties. Used for blankets, sweaters, pull overs, jackets, liners, jogging suits, hats and ski wear. Polar fleece can be made from recycled PET plastics such as soda bottles. The insulating properties of polar fleece compare favorably with the best types of wool, but polar fleece is flammable and that is something to keep in mind. Various colors of polar fleece are shown in the picture below.

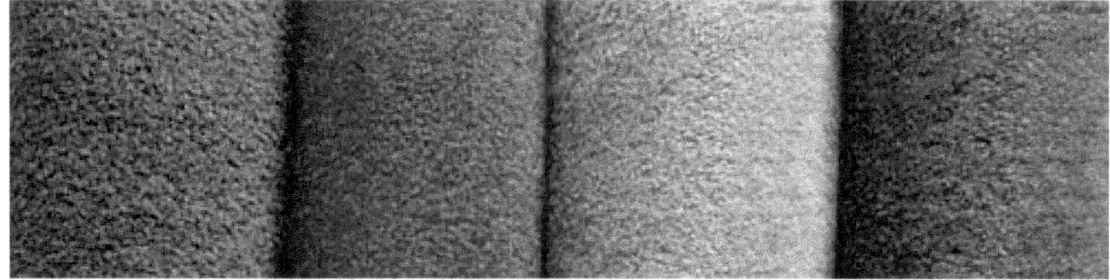

Microfiber polyester - Used in clothing, upholstery and many other applications, microfiber polyester is durable, cleans easily and is difficult to accidentally stain. Microfiber polyester can look like silk or fine cotton, but is much more resistant to wrinkling.

Microfiber ultrasuede - Is a non animal alternative to soft suede leather with improved durability and stain resistance. Looks just like suede leather, but is thinner, lighter and stronger.

Non-woven materials

Plastic leather - also called pleather or faux leather comes in many types. Some types have a woven cotton or polyester fabric backing. This non-animal alternative to leather is lighter than leather. Some newer types can outperform leather in some applications, but overall leather is more durable.

Felt - is produced by pressing fibers together (usually while wet). Felt comes in many types ranging from soft felt used as a lining material, to felt based construction materials that are used in structural applications. Felt is a popular material for making hats.

Fiber types

Polyester

- Polyester comes in both woven and knit fabrics is durable and relatively non-wrinkling. Absorbency is not that good, but once wet it dries quickly. Static electricity will occur in dry climates.
- It is the second most widely used fiber for clothing after cotton and is often blended with cotton in a 50/50 mix to combine the feel and absorbency of cotton to the durability and non wrinkling aspects of polyester.
- Polyester is made from synthetic fibers (thermoplastic) and is non-allergenic.
- Polyester is strong. Resistance to heat is OK, it gets sticky at 440°F and melts at 475°F. Resistance to abrasion is very good. Resistance to chemicals is excellent. Color fastness is good.
- UV resistance is very good. Polyester is preferred over nylon for outdoor applications with long term sun exposure.

Cotton

- Cotton is made from natural plant based fiber, takes dyes and printing inks well and can have vibrant colors. Not prone to static electricity.
- Cotton wrinkles and shrinks, it is often prewashed before sewing to reduce shrinking of clothing.
- Cotton feels good in contact with the skin and is the most used fabric for all types of clothing. It is absorbent and removes sweat from the skin which is good in warm climates. In cold climates cotton has inferior insulating properties compared to wool or synthetic polar fleece especially when wet.
- Cotton has lower strength compared to polyester and is often mixed with polyester to increase durability, decrease shrinking and reduce wrinkling.
- Cotton does not stretch very well, but has good heat resistance and can take higher temperatures than polyester. Cotton does not melt when ironed at high temperatures. Resistance to chemicals is not as good as polyester. Resistance to abrasion is not as good as polyester. Color-fastness is not as good as polyester. Cotton will sustain damage from continuous sun exposure.

Rayon

- Rayon is soft, absorbent and takes dyes, so it is a favorite for applications that require the most vibrant colors, high sheen and deep luster. Rayon feels good against the skin.
- Rayon wrinkles badly and is often mixed with other fibers to reduce wrinkling and increase durability. Rayon has low heat resistance and should be dried at low temperatures.
- Rayon is a semi-synthetic fiber that is made from highly processed natural fiber.
- Medium to low strength. Not as strong as polyester or cotton.
- Does not stretch well. Resistance to chemicals is not as good as polyester. Resistance to abrasion

is not as good as polyester or nylon. Color-fastness is not as good as polyester.

Nylon

- Nylon is good for jackets, rain wear, ski wear, backpacks, tents, luggage, cases, shoes, upholstery and other applications that require high strength and abrasion resistance. Not good for permanent sun exposure, UV resistance is OK but not good.
- Nylon is made from synthetic fibers (thermoplastic), and has high strength. Resistance to chemicals is good. Resistance to abrasion is very good. Resistance to heat is OK, it gets sticky at 440°F and melts at 495°F. Color fastness is good. UV resistance is good but not great. Polyester is preferred over nylon for outdoor applications with long term sun exposure.

Linen

- Linen is a natural fiber with properties similar to cotton, it is stronger than cotton but wrinkles more than cotton. Linen is used in shirts, jackets, slacks and hats.

Wool

- Wool is a natural fiber that comes from sheep. Cashmere and mohair are similar and come from goats. Alpaca and camel are similar and come from camel like animals and lamas.
- Wool has excellent insulating properties even when wet, so it is a favorite for cold weather clothing.
- Wool has good resistance to heat, good stain resistance, low to medium strength, high resistance to wrinkling, low resistance to chemicals. In general wool should be dry cleaned or hand washed and not machine washed or machine dried.

Silk

- Silk is a natural fiber made from the silkworm. Silk is used for shirts, suits, lingerie, bedding, ties, etc. Silk is strong, absorbent and feels good in contact with the skin.
- Silk has fair resistance to heat and should be ironed at low temperatures. Stain resistance is low, resistance to wrinkling is low, resistance to chemicals is low. In general silk should be dry cleaned or hand washed and not machine washed or machine dried.

Other fibers

- Natural fibers to check out are flax, jute, hemp and bamboo. These fibers will increase in usage in the future for environmental reasons.
- Synthetic fibers to check out are acrylic, olefin, modacrylic, carbon fiber, spandex, Nomex, Orlon, artificial silk, etc.
- New synthetic fibers are being developed that incorporate the desirable features of natural fibers with increased durability and strength.

Tips

Nap or pattern direction - Nap is the fuzzy surface of fabrics like velvet or felt, this surface is directionally oriented and will lay differently when smoothed depending which direction the fabric is facing. Some fabrics have patterns that are directionally oriented. Pay attention to nap and patterns when buying fabric and laying out and cutting fabric for items that have seams that must match up. If the fabric orientation is incorrect you may have to remove seams and re-cut fabric to fix it.

Quilting Machines

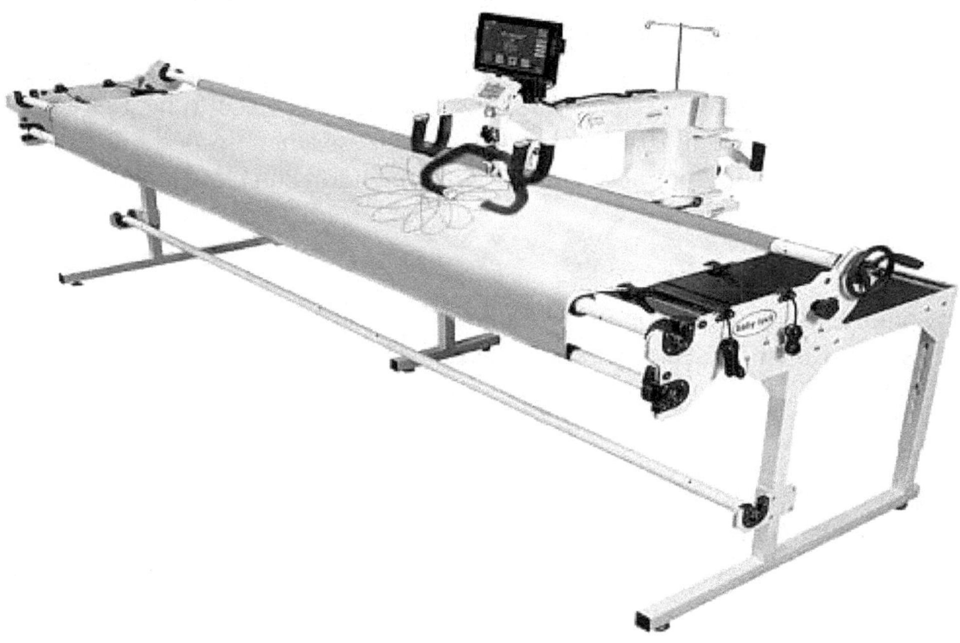

Quilting is the art of making quilts. Most quilts are made from 3 layers that are sandwiched together, the two outer layers are a light to medium weight fabric such as cotton or polyester (or a blend) and the center layer is a batting material or filler that is usually also cotton or polyester.

There are several operations that are done in traditional quiltmaking as follows:

Quilting - Sewing through all three layers to secure the layers together. This can be done in straight lines or in a pattern.

Edging (or binding) - Terminating the edges of the quilt is usually done by folding the outer layers at the edge and securing them with a straight stitch. Edging can also be done by sewing in a strip of bias cut fabric to cover the edges.

Piecing - Piecing is when the top or bottom layer is formed by joining many small pieces of fabric of different colors or patterns into squares (blocks). The blocks are then joined to form the layers. Popular block sizes are 9", 10" and 12". Piecing can also refer to assembling, in which the square blocks are sewn together to form the top or bottom layer of the quilt.

Assembling - The square blocks are sewn together to form the top or bottom layer of the quilt.

Free motion quilting - This is quilting that is done by manually controlling the feed of the fabric through the machine (similar to darning). With free motion quilting many complex and artistic patterns can be made because the fabric can be moved in any direction while sewing. To do free motion quilting the feed dogs of the machine must be disengaged (or covered with a feed dog cover plate) or a special quilting machine with no feed dogs must be used.

Most of the operations above are done with straight stitching, but some quilters also use zigzag stitching and decorative stitch patterns.

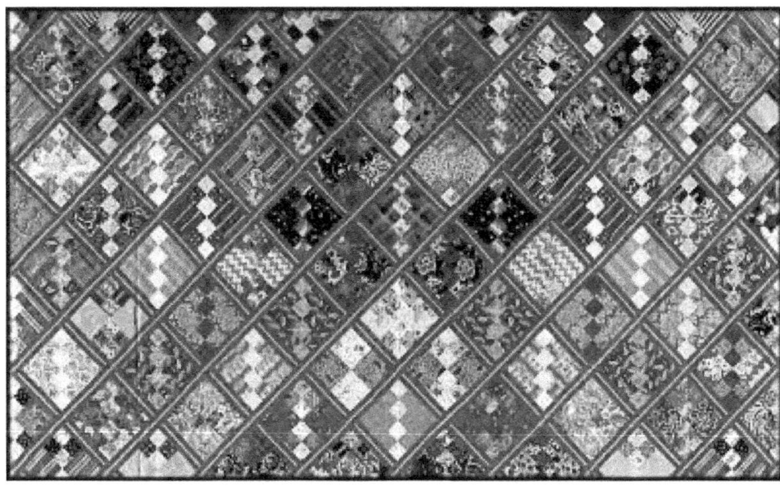

Types of machines and features

Straight stitch machines - Most quilters prefer a heavy duty straight stitch machine. Heavy duty straight stitch machines have less play (extra movement) in the needle bar (because there is no zigzag mechanism). This means that these machines are able to tolerate more pull on the needle from the fabric without miss-stitching or causing the needle to hit the needle plate. Sewing over seams and doing free motion work can pull the needle and these machines have a noticeable advantage in maintaining a consistent stitch under these conditions. Some quilters will also have a second machine to do zigzag stitching and decorative stitches.

Zigzag machines - Beginning quilters and quilters on a budget can use a basic zigzag machine for everything. Actually you can do most quilting on any type of machine, you just have to go slower and pay attention to make sure that you are not bending the needle.

Extended harp space - Machines with extended harp space and long beds are preferable for quilting because quilts are large and the extra space means you have more room to maneuver the quilt on the machine

Long arm machines and quilting frames - Specialized long arm quilting machines have no feed dogs and are made specifically for free motion quilting use in a quilting frame. A quilting frame is a large table-like rack that holds the quilt and has tracks for a long arm machine to mount on. Instead of the machine sitting stationary and the fabric moving like normal, with a quilting frame the quilt stays stationary and the long arm sewing machine rolls on the tracks (with wheels) to move around the fabric. Some setups even have handlebars so that the machine can be driven around the fabric like a motorcycle!

Stitch regulators - A stitch regulator is an electronic device used in free motion quilting to control the machine speed in response to the fabric movement. This allows the operator to freely move the fabric through the machine and the stitch regulator will speed up and slow down the machine as needed to make stitches of the desired length. Not all machines can be used with a stitch regulator. Some quilters prefer to control the machine speed themselves and do not like to use a stitch regulator, but other quilters really like them.

Industrial machines - Some quilters prefer to use a straight stitch drop-feed industrial machine for quilting. The advantage is that these machines are available in long arm versions for a very reasonable price. If you are interested in using an industrial machine for quilting make sure to get a servo motor that has good low speed characteristics (see the chapter on Industrial Machines).

Embroidery machines - Some quilters use embroidery machines to make designs for their quilting squares or top layers (see the chapter on Embroidery machines).

Small quilting frames - The above machines have about 7 inches of harp space. You can get small quilting frames for these machines and they are quite inexpensive compared to the big frames, but you will be limited to sewing smaller amounts of fabric at a time because of the limited harp space. If you want to work on larger areas of fabric you will have to get a long arm machine that has more harp space.

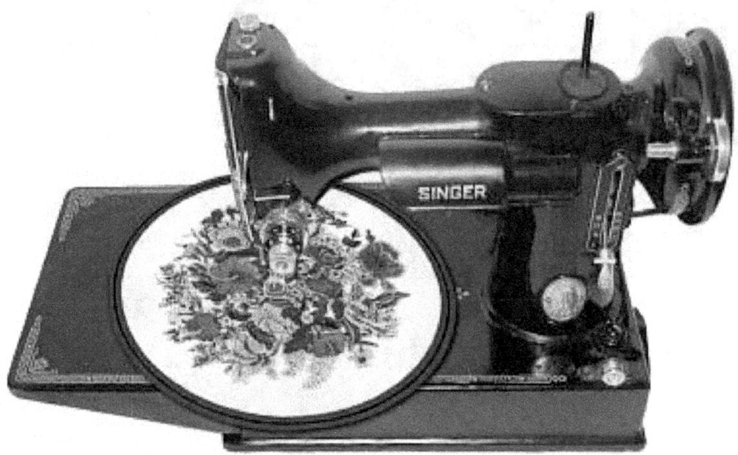

For small work and for sewing designs on quilting squares you can also use an embroidery hoop with one of the above machines set up for free motion work (feed dogs disengaged and using a darning or embroidery foot).

Inexpensive machines for quilting

Basic Zigzag - For quilting on a budget any basic zigzag machine can be used. If you are buying a new machine get a machine with a vertical oscillating hook and HA-1 type bobbin-case like one of the machines pictured below. If you want to do free motion work then make sure that the machine you are considering has drop feed dogs or that a feed dog cover plate is available for the machine. There are many models that will work fine, some examples are as follows:

Brother LS2125i

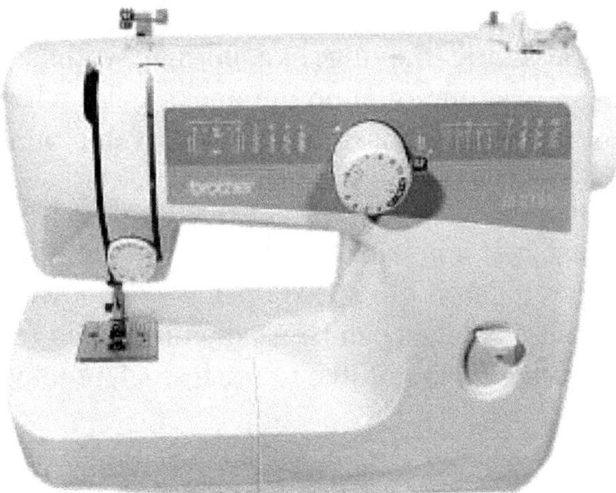

Brother LS2125

The Brother LS2125i sells for about $75 on Amazon and has a front-loading bobbin. Lacks a separate stitch length control.

Singer Tradition 2259

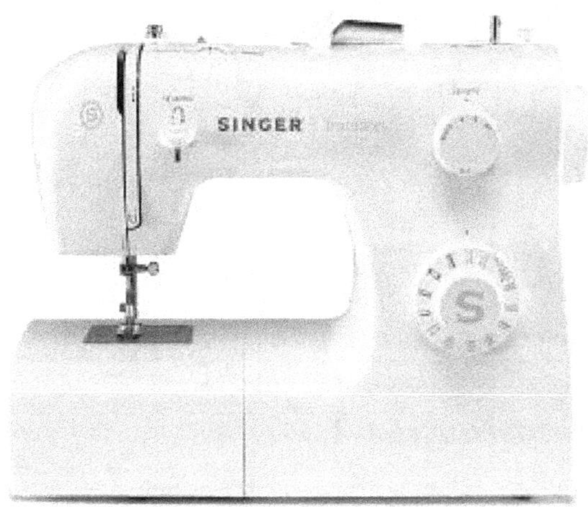

Singer Tradition 2259

The Singer Tradition 2259 sells for about $78 on Amazon and has a front-loading bobbin. Has a separate stitch length control.

Vintage Heavy-Duty Straight Stitch machines are favorites with quilters on a budget. The most popular versions are as follows:

Singer 15-91

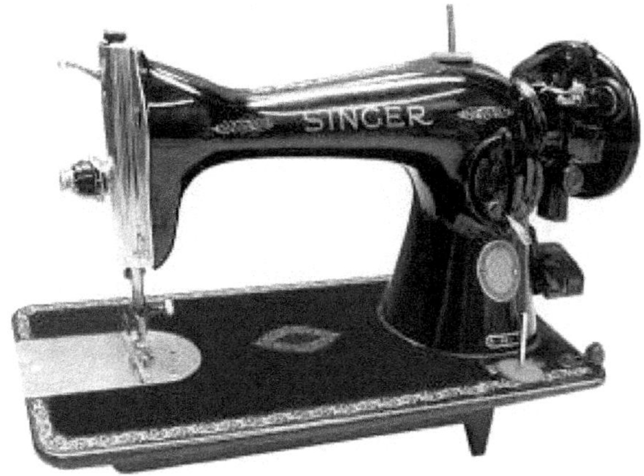

Singer 15-91

The Singer 15-91 sells for about $50 to $200 on Craigslist. These straight stitch machines take needles up to size 21, They the toughest machines you can get without going to an industrial machine.

Japanese HA-1

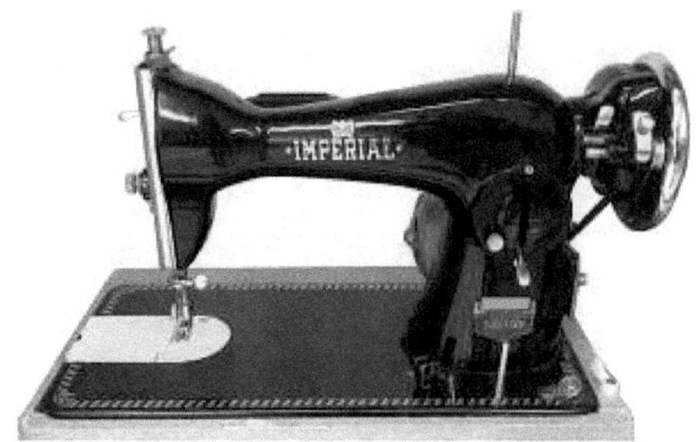

Imperial HA-1

The Japanese HA-1 machines sell for about $50 to $120 on Craigslist. These are belt drive machines similar to the Singer 15-91, some people prefer the belt drive over the gear drive in the 15-91.

Medium price machines for quilting

Additional harp space, high speed and a thread cutter really are nice features to have for quilting. The following machines have these features and are excellent for quilting. These machines are powerful, fast and very smooth. They have excellent low speed control and sew through thick materials with ease. They have needle up/down controls, knee lifters, automatic thread cutters and long beds with 8.5 inch harp space. There is a good selection of reasonably priced free motion quilting frames made for these machines.

Brother PQ-1500s

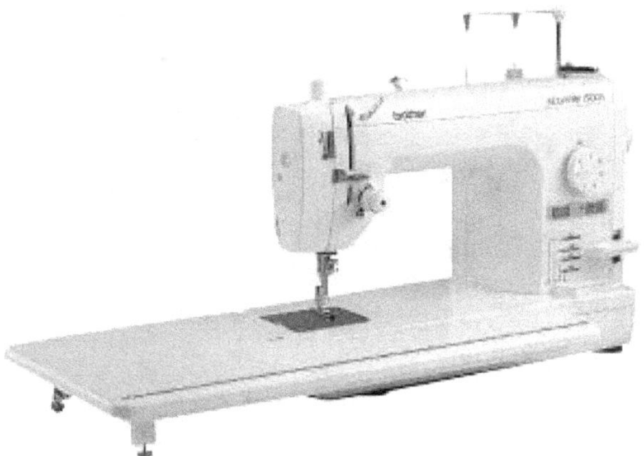

Brother PQ-1500S

The Brother PQ-1500s sells for about $669 on Amazon and has pin feed for hard to feed fabrics.

Juki TL-2010Q

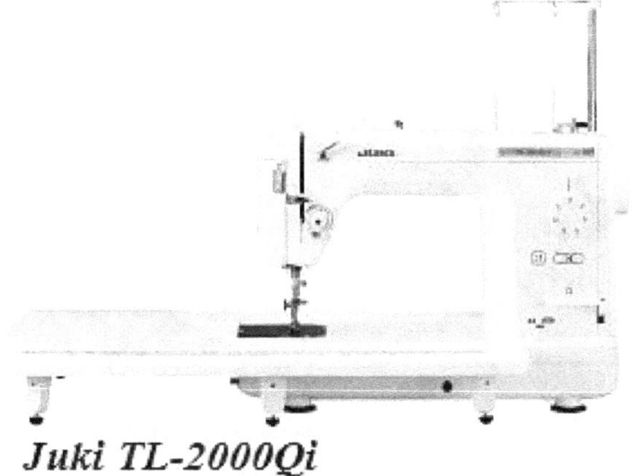

Juki TL-2000Qi

The Juki TL-2010Q sells for about $849 through on-line stores. It is a heavy-duty electronic high speed straight stitch machine.

Janome 1600P-QC

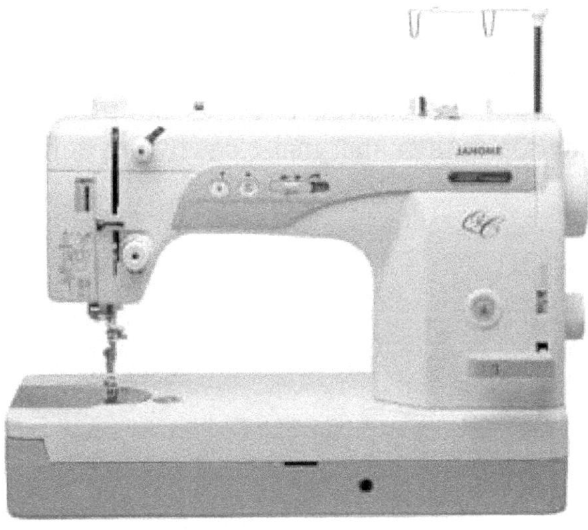

Janome 1600P-QC

The Janome 1600P-QC sells for about $799 through on-line stores. It is a heavy-duty electronic high speed straight stitch machine.

Long arm machines for quilting frames

If you want a long arm machine designed for free motion quilting such as the one from Baby Lock pictured below you should check out the one of the specialty long arm quilting machine suppliers. These machines range from about $2500 (on sale) to over $10,000 depending on the model and features. A few brands that are popular in the US are as follows:

- Handi Quilter
- TinLizzy
- Baby Lock
- Grammill
- Nolting

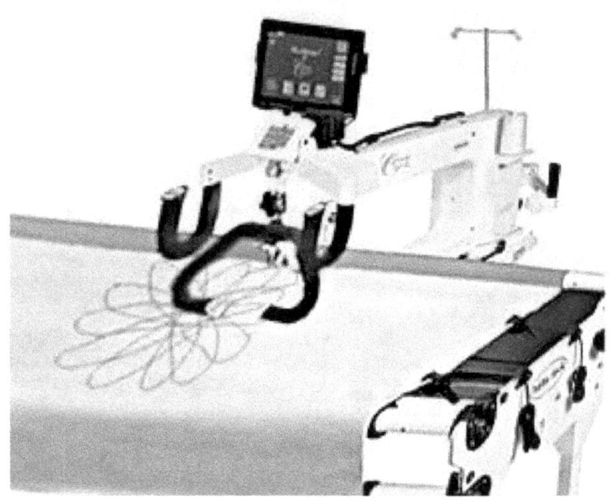

Most long arm quiting machines function in basically the same way. Some models have features that others do not such as built in stitch regulation. It is a good idea to go to a dealer that is able to let you test out different models because some models are great for one style of sewing, but may not be so good for another style.

You should not buy an expensive long arm machine and quilting frame before you spend some time free motion quilting on a small machine or using a rental machine to gain experience and to know what features you need. I have seen many expensive long arm quilting machines sitting in houses for years with dust collecting on them because the purchase seemed like a good idea at the time, but the machine did not end up working as originally envisioned. Beautiful freestyle and pattern work can be done with a long arm machine, but there are many top quilters that make patchwork quilts and do not need or use a long arm machine.

Embroidery Machines

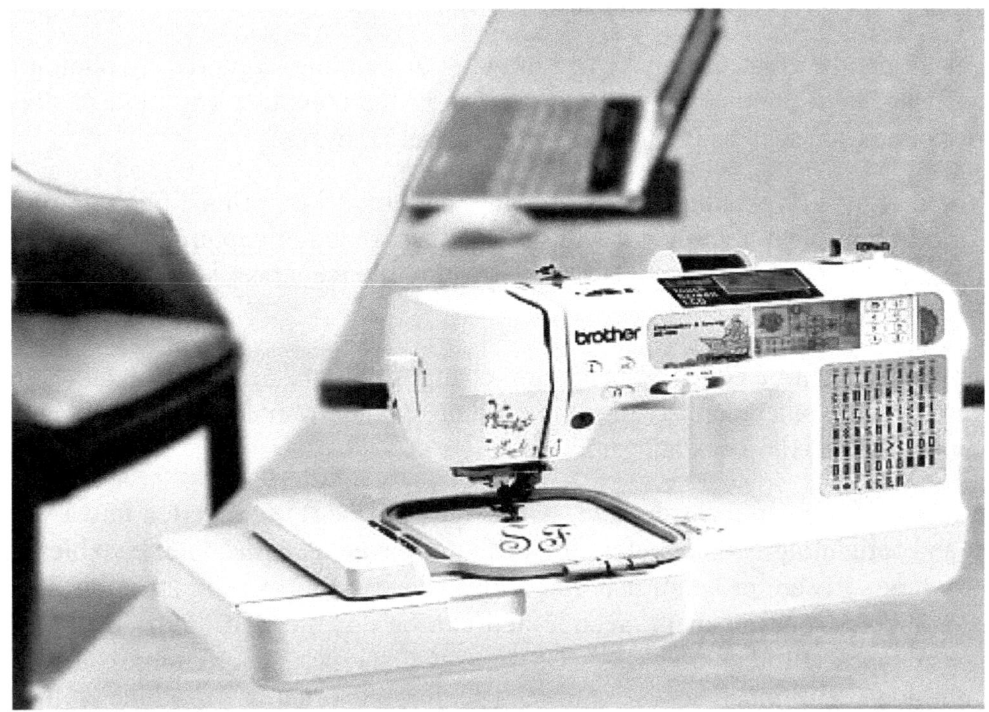

Machine embroidery can be done with regular sewing machines using free motion techniques, but most people associate free motion with quilting. Machine embroidery most commonly refers to computer controlled embroidery that is done by a sewing machine under computer control. That is the type of embroidery we will be covering in this chapter.

Most embroidery machines are sewing machines that have a built in computer that controls the machine and two additional motors that drive a hoop. The fabric is inserted into the hoop and the hoop is snapped into the drive assembly on the the base of the machine. The machine can move the hoop in any direction while sewing to produce the desired embroidery patterns.

Embroidery machines function in a similar manner to a computer printer. The content (in this case the embroidery design) comes from an external source (such as a personal computer) and is transferred to the embroidery machine using a memory card or USB memory stick. Alternatively some embroidery machines can receive files from a computer through a USB cable or other type of computer interface cable. Once the file is transferred to the embroidery machine the machine can sew the design onto the fabric.

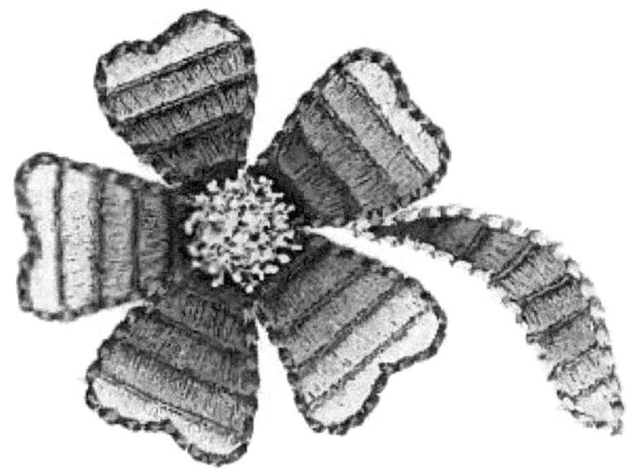

Work flow and software

Work flow is the combination of steps and processes needed to go from an idea to the final output (in this case a finished design embroidered onto a piece of fabric).

Design - The design can be created on paper by drawing or painting, captured by photograph, or can be created using software in a computer. To create designs in the computer any kind of drawing or photo editing software can be used such as Photoshop, Gimp, Corel Draw, etc.

Digitizing - If the design is originating from a drawing, painting, film, printed photograph, existing piece of embroidery or other physical item then it will need to be scanned or captured using a flatbed computer scanner or digital camera. After the design is scanned you will have a raster image file (bitmap file). This is usually in jpeg or png format.

Editing - The image file can be edited to make any changes that are needed such as changing colors or removing backgrounds or other parts of the image that are not wanted. Some popular image editing software that can be used is Gimp or Photoshop.

Converting and formatting - In this step the raster image file (jpeg) is converted into a vector image and then processing and formatting is done before the design is saved as a machine readable embroidery file. A popular embroidery software program that does converting and formatting is Embird. There are many other programs available, some of them are free and can be downloaded. Some of the file formats for embroidery files are .pes (Brother and Baby Lock), .jef (Janome), .dst (Tajima), .hus (Viking), .vp3 (Pfaff). The embroidery file must be the correct type for the brand of machine you are using. The formatting steps can include re-sizing, rotating, changing the colors and editing the actual stitch coordinate data.

Transfer the design to the embroidery machine - The embroidery file must be transferred from your computer to the embroidery machine. Some embroidery machines use a memory card, others have a computer interface such as USB or Ethernet.

Set up the embroidery machine - The fabric is loaded into the hoop and stabilizer is add if it is needed and the machine is threaded with the correct color thread (or threads). Stabilizer is a thicker material that is added under the fabric to keep it from stretching or "becoming dimensionally unstable" as they like to say. If the fabric stretches during embroidery it can result in ugly miss-stitching and jamming of the hook.

Run the job - The machine is run. The operator should monitor the machine as it progresses to make sure that nothing goes wrong and to stop the machine if there is a problem like thread breakage or jamming. Depending on the machine and the design, the machine may stop during the job and need to have the thread (or threads) changed to a different color.

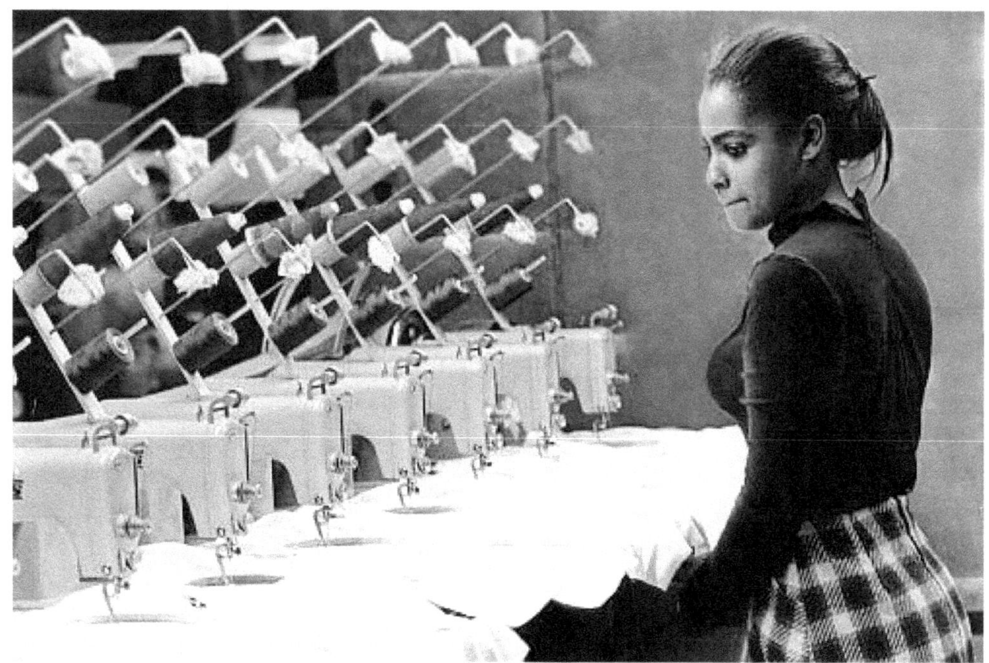

Features and specifications

Editing and design capability - Some embroidery machines have built in LCD touch screens and are capable of editing designs before they are sewn or even creating new designs right on the touch screen of the machine. The editing and design capabilities that are built in to these machines can be convenient, but are simplistic compared to computer software that does the same thing.

Built in memory - Most embroidery machines have built in memory and can hold a number of designs in memory at a time. Some machines come with design libraries in permanent memory.

Multi-needle - Two or more needles are used so that the machine can have multiple thread colors loaded at the same time. There are some machines with 6 or more needles that can run complex designs without needing to stop for thread changes.

Image sensors or cameras - Some higher end machines have special sensors or cameras and imaging software that watch the needles as they are sewing and can recognize and even correct for some errors.

Fonts and monograms - Many machines have built in fonts and monograms so that they can produce simple designs (monograms and name tags) without a computer.

Needle threader, automatic thread cutter, etc - Almost all machines have these.

Speed control - Lets the user set the machine to a faster of slower speed depending on the type of fabric and thread being used and the type and complexity of the design.

Hoop size - Hoops come in many sizes and each machine has a maximum hoop size. Inexpensive machines start with a 100mm x 100mm hoop size, medium home machines have a 150mm x 200mm hoop size and commercial machines can have much larger hoop sizes.

Inexpensive embroidery machines

Inexpensive machines have small hoop sizes and are good for smaller size projects. They are a great way to get started if you are interested in machine embroidery. Some inexpensive machines can also function as a regular sewing machine.

Brother PE500

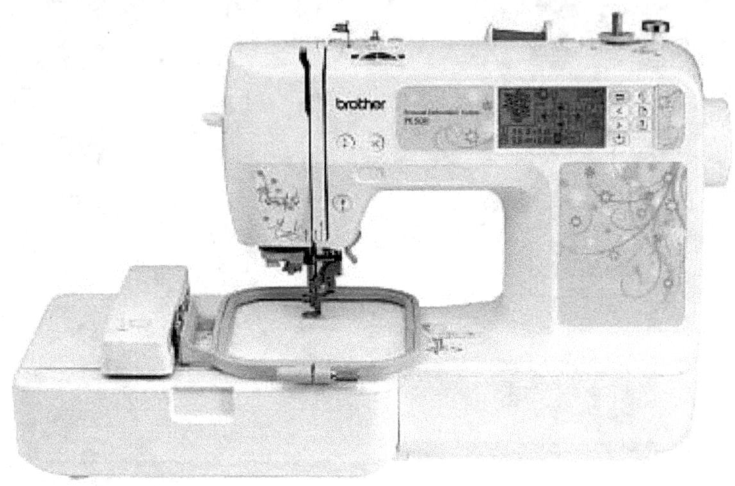

The Brother PE500 sells for about $269 on Amazon and has a thread cutter and max speed of 400 SPM. The maximum hoop size is 100mm x 100mm. It has 5 built in letter fonts and does monograms.

Brother SE400

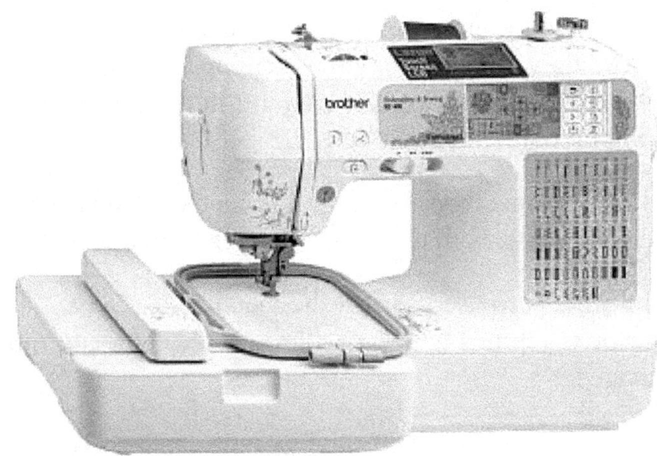

The Brother SE400 sells for about $337 on Amazon and also functions as a sewing machine. Max speed for sewing is 710 SPM and 400 SPM for embroidery. It has as a thread cutter and needle position control. The maximum hoop size is 100mm x 100mm. It has 5 built in letter fonts and does monograms. Includes a USB computer interface.

Janome MC 200E

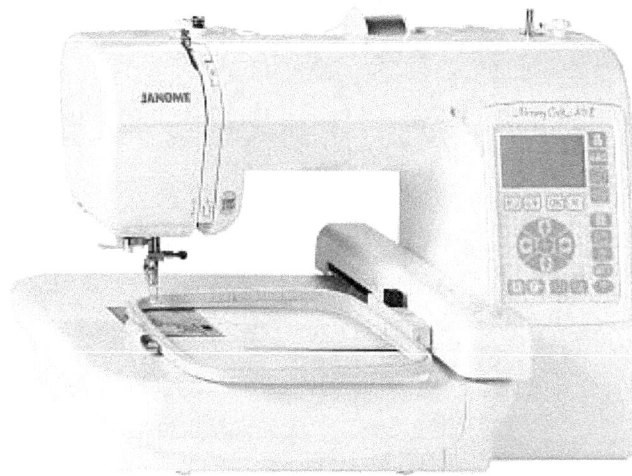

The Janome MC 200E sells for about $499 on Amazon and has as a thread cutter. The maximum hoop size is 125mm x 125mm. It has 3 built in monograms fonts. Includes a USB computer interface. Maximum speed is 650 SPM. Internal memory can hold 100 designs.

Brother PE770

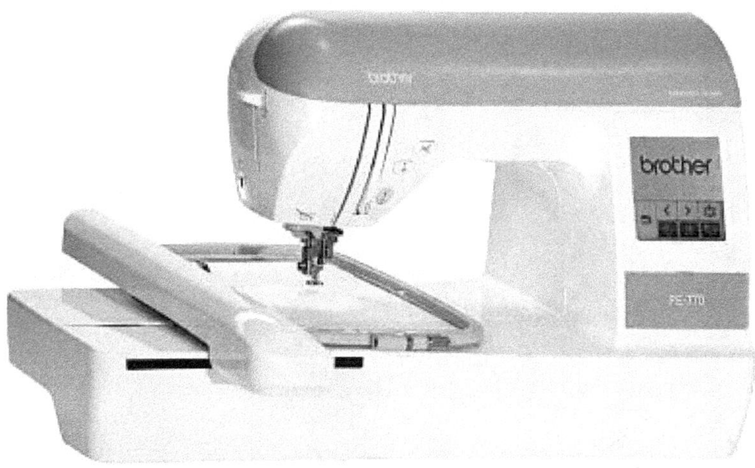

The Brother PE770 sells for about $590 on Amazon and has a thread cutter and max speed of 650 SPM. The maximum hoop size is 125mm x 250mm (with optional large hoop). It has 6 built in letter fonts and does monograms. USB computer interface. Some built-in design editing capabilities.

Medium cost embroidery machines

Medium cost machines are priced in the range of $1200 to $2500. I will not list prices for these machines because there are many promotional sales events and discounts available and the street prices are always changing, do your homework and wait for a low price if you are in the market for one of these machines.

Brother Innov-is 1500D

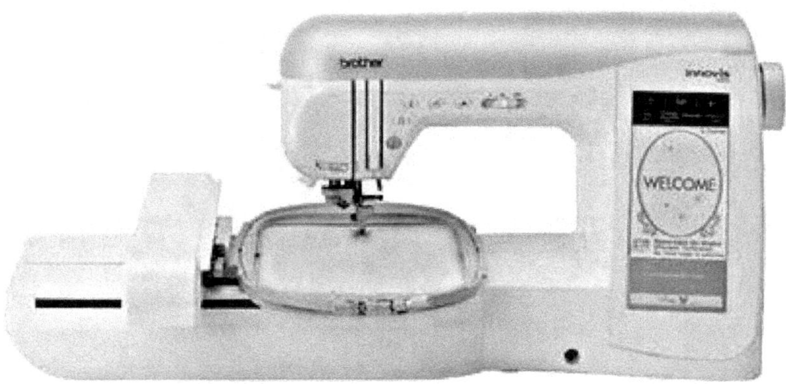

The Brother Innov-is 1500D does both sewing and embroidery and has a thread cutter and max speed of 1000 SPM for sewing and 800 SPM for embroidery. The maximum hoop size is 250mm x 150mm. It has many built in letter fonts and does monograms. USB computer interface. LCD touch screen.

Brother DreamMaker XE VE2200

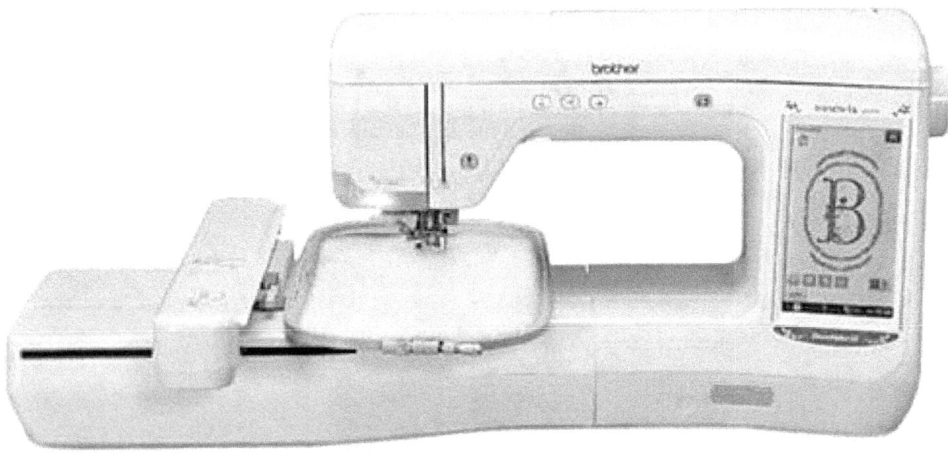

The Brother DreamMaker XE VE2200 has a max speed is 1050 SPM. It has as a thread cutter and needle position control. The maximum hoop size is 175mm x 300mm. Includes a USB computer interface.

Janome MC 350E

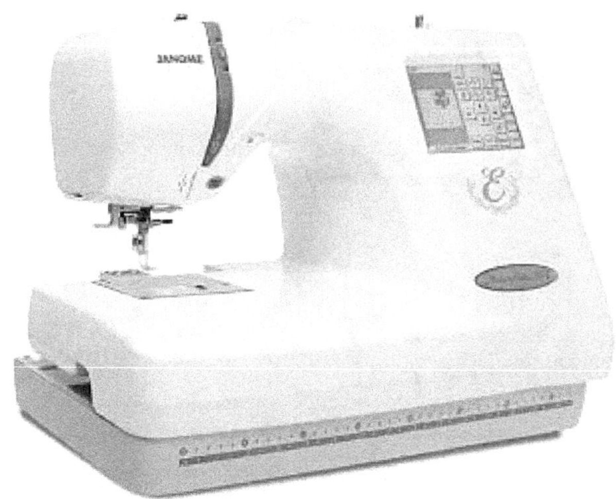

The Janome MC 350E has as a thread cutter. The maximum hoop size is 140mm x 200mm. It has 3 built in monograms fonts. Includes a USB computer interface. Maximum speed is 650 SPM. Internal memory has 100 designs.

Janome MC 9700

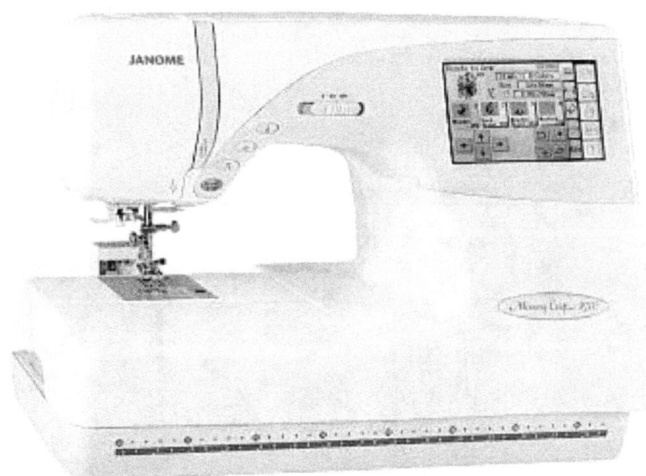

The Janome MC 9700 does both sewing and embroidery and has as a thread cutter. The maximum hoop size is 140mm x 200mm. It has 3 built in monograms fonts. Includes a USB computer interface. Maximum speed is 650 SPM. Internal memory comes with 95 designs.

Multi-needle embroidery machines

The multi-needle machines listed below are priced in the range of $5000 to $13000. There are also multi-needle multi-head machines available with from 2 to 15 heads (or more). The multi-head machines can cost over $25,000 and are for production use.

Janome MB-4

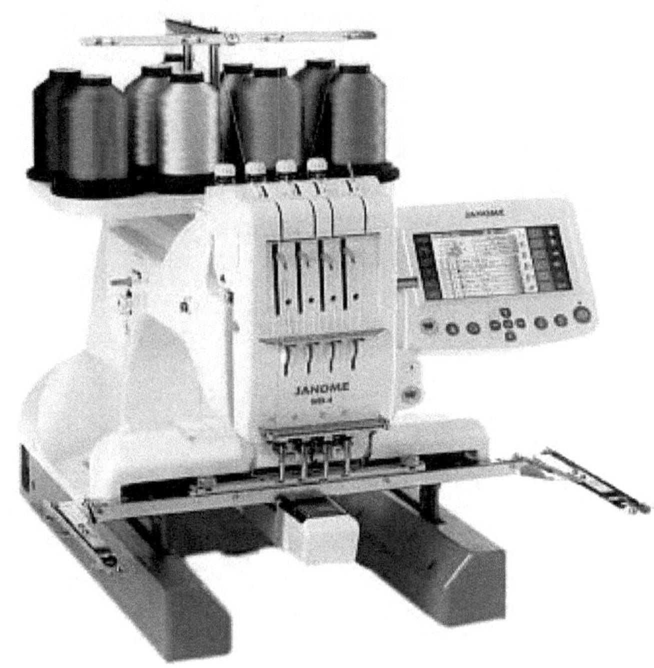

The Janome MB-4 has 4 needles. The maximum hoop size is 240mm x 200mm. It has 10 built in monograms fonts. Includes a USB computer interface. Maximum speed is 800 SPM. Internal memory has 50 designs.

Brother Entrepreneur PR650e

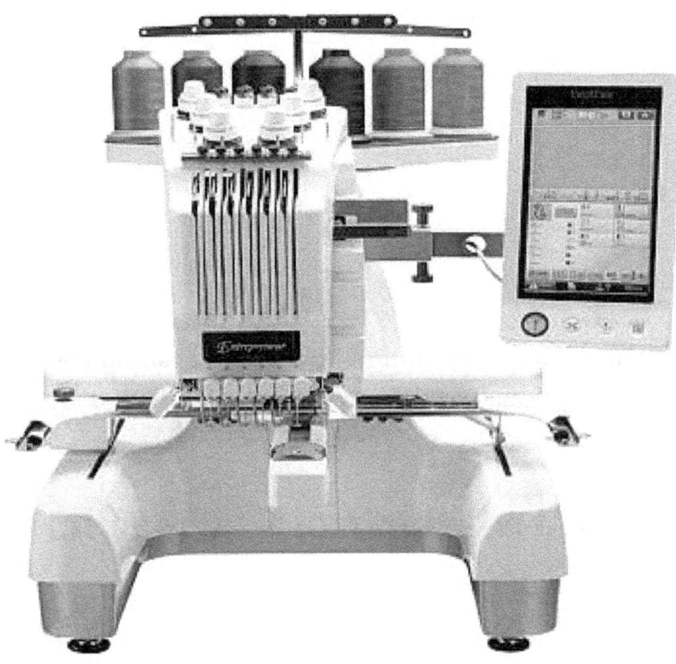

The Brother Entrepreneur PR650e has 6 needles and max speed of 1000 SPM and automatic thread cutters on all 6 needles. The maximum hoop size is 300mm x 200mm. It has a 300 color palette, 3 USB computer interfaces. LCD touch screen.

Brother Entrepreneur PR1000e

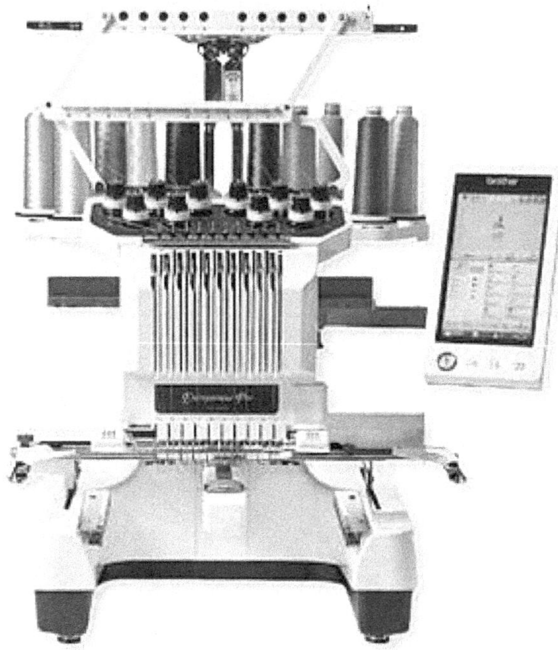

The Brother Entrepreneur PR1000e has 10 needles and max speed of 1000 SPM and automatic thread cutters on all 10 needles. The maximum hoop size is 350mm x 200mm. It has a 300 color palette, 3 USB computer interfaces. LCD touch screen. Built in machine vision to aid in needle positioning.

Melco Amaya Bravo

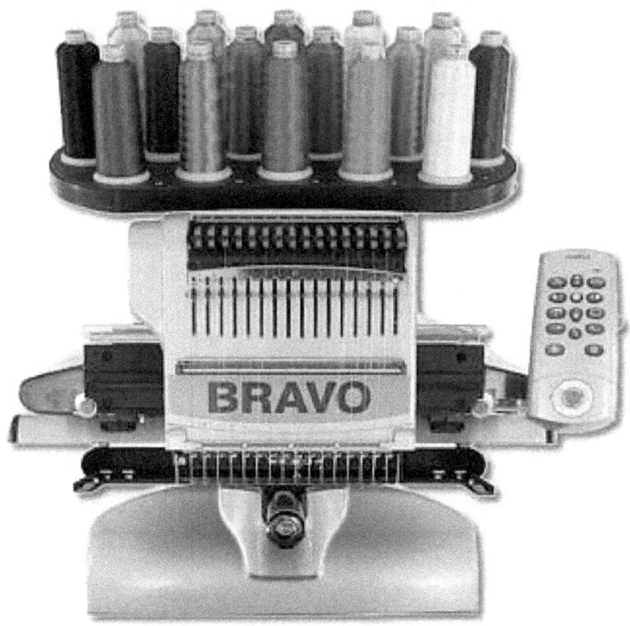

The Melco Amaya Bravo is made in USA and has 16 needles. The maximum hoop size is 360mm x 300mm. Includes a USB computer interface. Maximum speed is 1000 SPM. Internal memory comes with 95 designs.

Troubleshooting

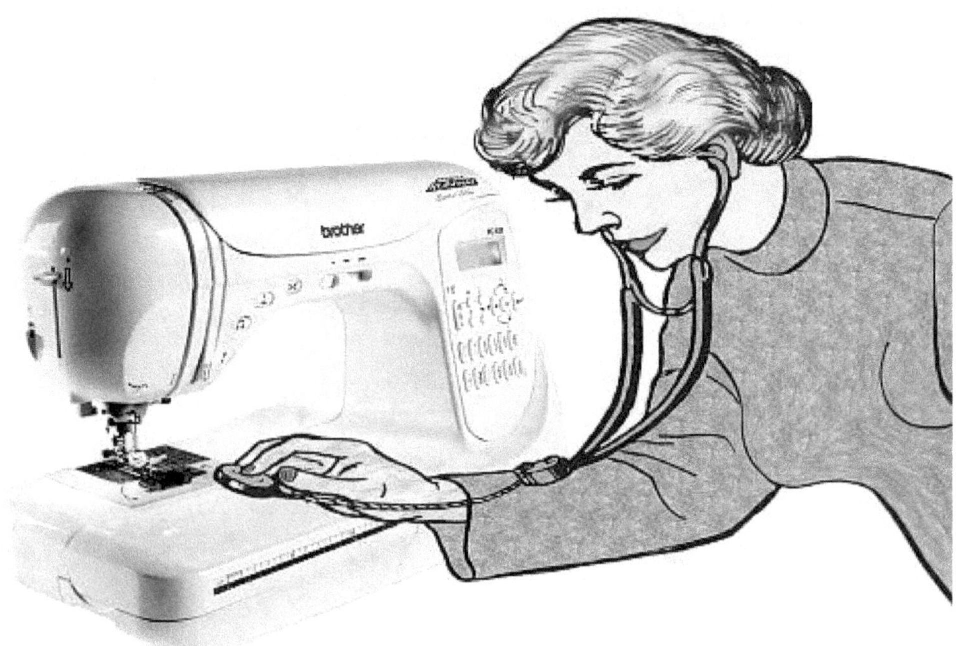

Troubleshooting by symptom

This section is for troubleshooting single problems in an otherwise well running machine. If your machine has multiple problems or needs serious help and you want to tackle the problem yourself then get a copy of our book "The Sewing Machine Master Guide" which has in depth sections on troubleshooting, adjustment and repair. Otherwise consider taking your machine to a repair shop.

This chapter has several troubleshooting checklists, go to the appropriate checklist that matches the symptom you are having.

- Needle thread breaks or stitch problems
- Bobbin thread breaks or stitch problems
- Needle breakage
- Fabric bunching
- Loose stitching
- Machine runs slow or is noisy
- Machine does not run.

NOTE - Keep in mind that the pictures of the thread path for the machines pictured in this chapter may not be the same as your machine. It is imoprtant to also refer to the diagrams in the owners manual for your specific machine.

Needle thread breaks or stitch problems

1. Threading and thread guides - Is the machine threaded correctly? Try re-threading the machine. Compare the way your machine is threaded to the threading diagram in the owners manual to make sure the threading is correct. Check to see that the thread guides are clean and that the thread can pull smoothly from the thread spool. Make sure that the thread is not stuck or snagged on any of the thread guides or the thread spool. The picture below shows the thread path of a common modern machine.

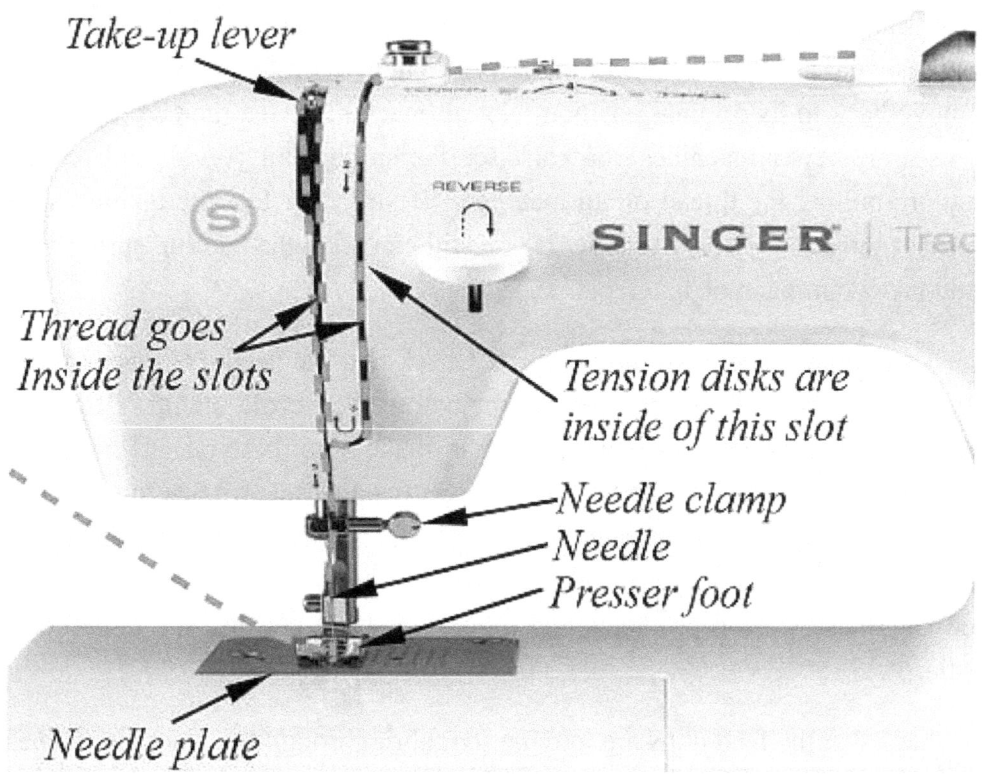

2. Tension disks - To check if the thread is feeding smoothly through the tension disks try the following test; With the presser foot down and the thread tension set to a normal setting (3 or 4 on most machines), un-thread the needle and pull some thread out from the take-up lever (this is shown in the picture below). Does the thread pull smoothly through the tension disks but with some tension? If not you should figure out why and correct the problem. It may be that the thread is hanging on the thread spool or thread guides, you may try cleaning the tension disks and thread path.

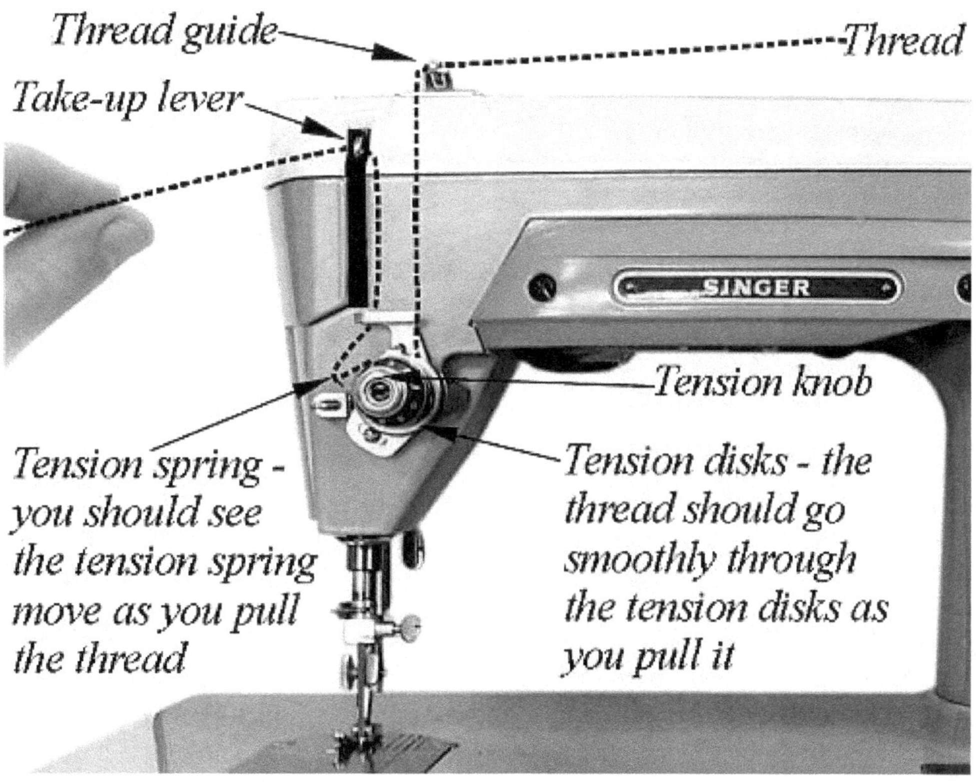

3. take-up spring - Pull the thread from the take-up lever with your fingers (as shown in the picture above). When you start to pull the thread you should feel the take-up spring pulling against you. As you pull the thread you should see the take-up spring move. When you release the thread you should see the take-up spring move back to its original position. Not all machines have the take-up spring in a location that is visible, on some newer machines you can't see the spring, but you should feel the force of the spring while you are pulling the thread on all machines. If you can't feel the take-up spring you should first check for proper threading of the machine. If you still can't feel the take-up spring when you pull the thread you should take your machine to a repair shop.

4. Needle - Check that the needle is not dull or bent. Check that the needle is inserted correctly and that the needle is inserted all the way into the needle bar before the needle clamp is tightened. Try a new needle. Try a new needle from another manufacturer to make sure that you did not get a bad batch of needles. Is the needle the correct size and from the correct needle system? See the chapter "Needles" for an explanation of needle systems. Your owners manual will list the correct needle system and size of needle for your machine. Make sure your needle is not too small, using a small needle with medium or large size thread can cause tension problems and thread breakage because the thread can not easily go through the needle.

5. Feed dogs - Check that the feed dogs are moving the fabric through the machine. Is the fabric moving forward? If not then try setting the machine to a medium or long stitch length. Very short stitch lengths can cause feed problems with some fabrics. Is there adequate presser foot pressure to hold the fabric in contact with the feed dogs? You may need to adjust the presser foot pressure if your machine has adjustable presser foot pressure. If the fabric is not moving through the machine try running the machine with no thread and see if the fabric will move through the machine. Sometimes tangled thread can stop the fabric from moving through the machine.

Check that the feed dogs are moving. To do this first un-thread the machine and turn the handwheel so that the needle is in the raised position. The feed dogs should also be in the raised position like in the first picture below.

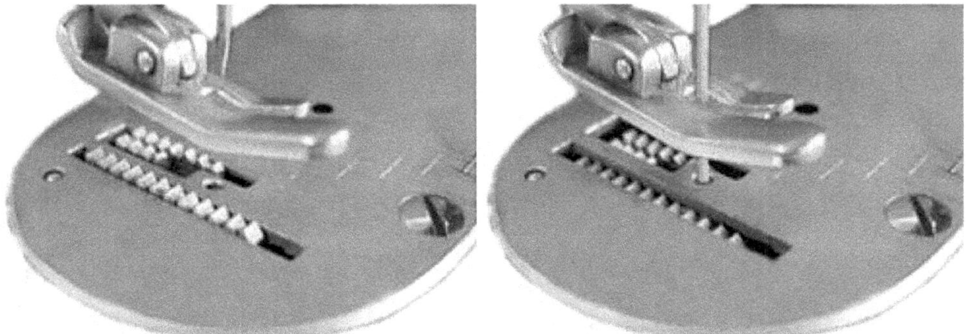

As you continue to turn the handwheel the feed dogs should move to the back of the machine and then as the needle moves down the feed dogs should also move down below the needle plate as in the second picture.

If the feed dogs are always in the lower position, check to see if your machine has dropping feed dogs. There may be a feed dog dropping knob and it may be in the down position. Check your owners manual, some machines have the feed dog dropping knob in a strange location like under the machine or in the back of the machine. If your machine does not have a feed dog dropping knob or if it does, but the knob is not in the dropped position and the feed dogs are still too low to move the fabric then your machine has a problem and needs to go to the shop for adjustment or repair. If the feed dogs are not moving at all your machine also needs repair.

6. Starting tension - Some machines do better if you hold the ends of the thread for the first few stitches while starting a seam to provide a little starting tension. This is only works with some machines and is described in more detail in the chapter "How to use a sewing machine".

7. Thick seams - Does the thread break when sewing thick seams but otherwise the machine sews OK? Is the fabric feeding through the machine correctly on thick seams? If not then make sure that you are not pushing or pulling the fabric through the machine, the feed dogs have to do the work. Some seams are too thick for the feed dogs to handle. If you are trying to sew a thick seam and the feed dogs are not able to move the fabric over the seam there is a way to manually advance the machine, see the section "Tips and ideas" in the chapter "How to use a sewing machine". Make sure that you are not trying to sew seams that are too thick for the type of machine you are using. With most home machines that would be about 6 layers of denim and with heavy duty machines 8 layers of denim.

8. Tension release - With the needle still unthreaded (like we did for step 2 and 3 above) move the presser foot to the up position and pull the thread again. Does the thread pull more easily then with the presser foot down? If not then your machine has a broken or miss-adjusted tension release mechanism and should go to the repair shop. Some very old vintage machines do not have a tension release, in that case ignore this step.

9. Needle hole - Is the needle hole obstructed or damaged? Check to see that a needle did not hit and damage the needle plate around the needle hole. You may need a flashlight and magnifier to properly inspect the needle hole.

10. Hook-race area - Is the hook-race area clean and free from thread knots, broken pieces of thread and lint? Is the hook-race properly oiled? (only if your machine needs oil in the hook-race, some machines do not require oil, check your owners manual).

11. Knotted or tangled thread - Check for thread that is knotted, has tangles or is sticking on the spool. Knotted or tangled thread can hang in the thread guides, tension disks or the needle and cause stitch problems and tread breakage.

12. Bad thread - Check for bad or rotted thread. Try breaking some of the thread in your hands. Compare to other thread if you are not sure. Old, damp or bad thread can rot or have other problems. Thread can be improperly manufactured or improperly labeled (such as thread with the wrong twist).

13. Excessive thread tension - Are you using too high a thread tension for the size of thread you are using? Try setting your thread tension to the normal setting for your machine. See the chapter on adjusting tension. If there is excessive thread tension but your tension knob is set to a normal setting then you may have a problem with your tension assembly. Try cleaning the tension disks.

14. Balanced stitch - Check to see of the stitch is balanced. If the lock point is not in the center of the stitch then adjust the thread tension so that the stitch is balanced (see the chapter Adjusting Tension).

Could not find the problem? If you are still experiencing problems then continue with the next section "Bobbin thread breaks and skipped stitches".

Bobbin thread breaks or stitch problems

1. Bobbin check - Replace the bobbin with another bobbin. Make sure the bobbin is not over filled. Make sure that the bobbin is the correct size and type for your machine and that the bobbin is inserted correctly into the bobbin-case. Look closely at the bobbin to make sure that it is not warped or damaged.

Bobbin Problems:

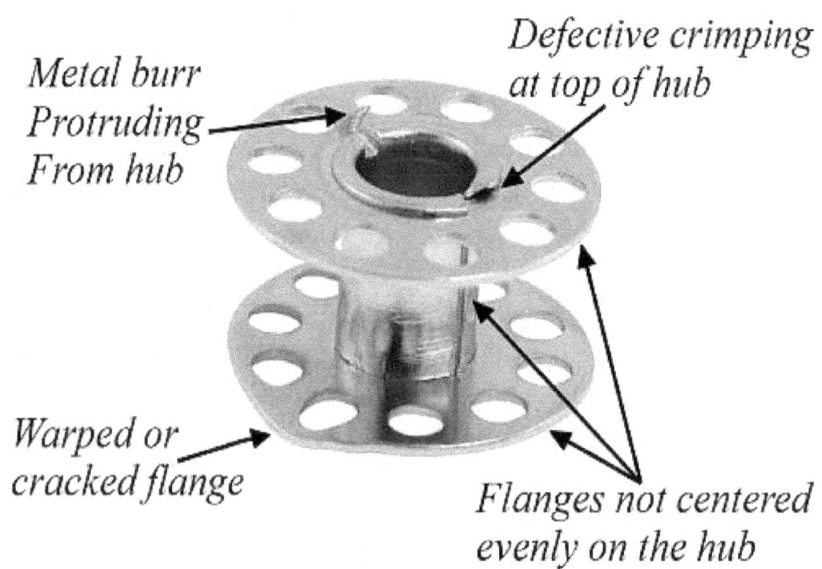

Make sure that the bobbin is not over filled. With an over filled bobbin the thread can contact the inner wall of the bobbin case. This will stop the bobbin from turning freely and cause stitch problems or broken thread.

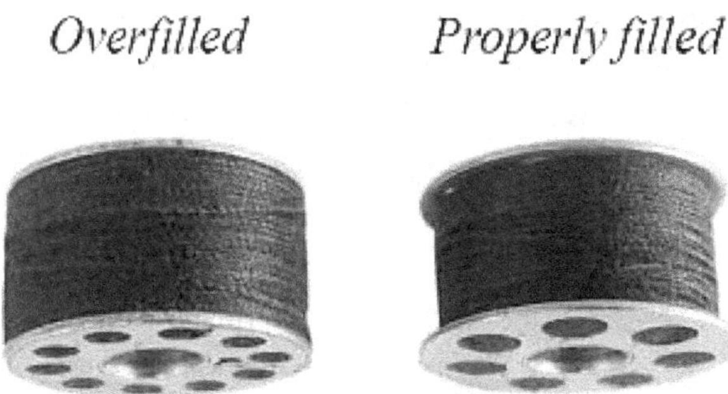

2. Bobbin-case and bobbin threading - Inspect the bobbin-case and make sure that it is clean. Make sure that there are no pieces of loose thread or lint in the bobbin-case and bobbin tension spring area. Check to see that the bobbin-case is threaded correctly and that the thread is going through the bobbin tension spring.

3. Bobbin tension - As pictured below slowly pull a few feet of bobbin thread from the needle hole. Make sure that you are pulling the bobbin thread and not the needle thread. Also make sure that the bobbin thread is not pinned under the presser foot, the thread should go in between the toes of the presser foot or the presser foot can be in the up position. Does the thread pull smoothly from the needle hole with some tension? If not then check the bobbin-case and bobbin tension spring. If it is hard to pull the thread then check to make sure that the bobbin is not over filled or that there is not too much bobbin tension (on machines with adjustable bobbin tension). Try using another bobbin, you may have a bad bobbin.

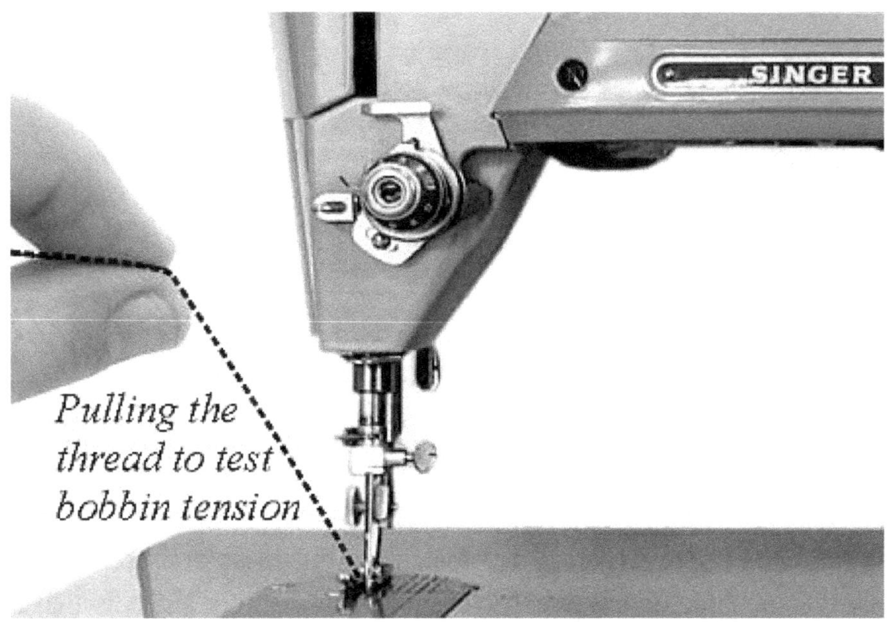

4. Needle - Make sure that the needle is not dull, damaged or bent. Is the needle inserted all the way into the needle bar before being tightened? Try a new needle and make sure it is inserted properly. Try a new needle from another manufacturer to make sure that you did not get a bad batch of needles. Is the needle the correct size and from the correct needle system? (see the chapter Needles).

5. Bad thread - Check for bad or rotted bobbin thread. Try breaking some of the thread in your hands, is it as strong as it should be? Compare to other thread if you are not sure.

6. Needle hole - Is the needle hole obstructed or damaged? Check to see that a needle did not hit and damage the needle plate around the needle hole. You may need a flashlight and magnifier to properly inspect the needle hole.

7. Hook race - Is the hook area clean? Does your machine require oil in the hook race? See the section "Hook and feed dog area" in the chapter "Maintenance".

Could not find the problem? If none of the above checks isolate the problem you may want to take your machine to a service shop. If you want to try to repair the machine yourself then get our other book, "The Sewing Machine Master Guide".

Needle breakage

1. Needle - Are your needles OK and not bent or defective? Test the needle (see the section “Inspecting needles” in chapter “Needles”) or try a new needle from another manufacturer to make sure that you did not get a bad batch of needles. Is the needle the correct size and from the correct needle system? (see the chapter Needles).

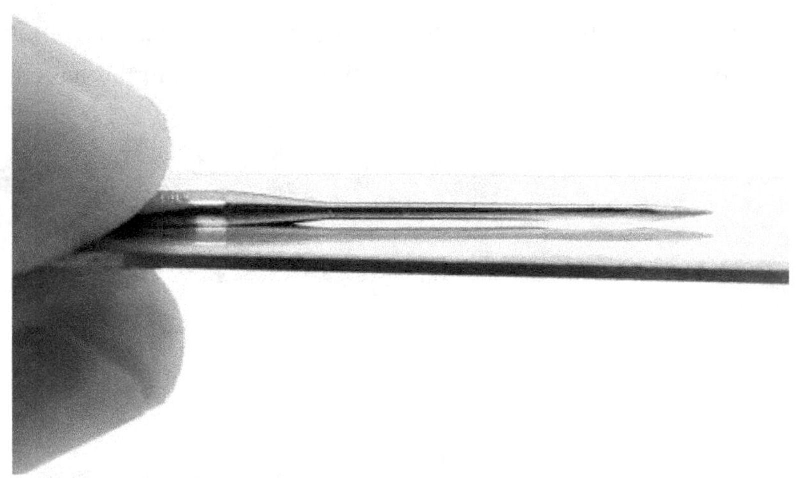

This needle is straight

2. Needle insertion - Is the needle inserted properly? Is it inserted all the way into the needle bar before being tightened? Is the needle clamp or screw tight?

3. Needle size - Is the needle size too small for the fabric thickness you are sewing or too small for sewing through thick seams? Smaller size needles are not as strong. Try a needle that is one or two sizes larger if your machine will take a larger size.

4. Thick fabric or thick seams - Does the needle break when sewing thick fabric or thick seams? Is the fabric feeding through the machine correctly on thick seams? Make sure that you are not pushing or pulling the fabric through the machine, the feed dogs must to do the work. If you push or pull the fabric it will bend the needle and the needle will hit the needle plate and break. The feed dogs move the fabric only when the needle is in the up position, they do not move the fabric when the needle is down. If you push or pull the fabric through the machine you will bend and break the needle. See the chapter "How to use a sewing machine" for instructions on how to sew thick fabric or thick seams.

5. Feed dogs - Check that the feed dogs are moving the fabric through the machine. Is the fabric moving forward? If not then try setting the machine to a medium or long stitch length. If the stitch is too short or if the fabric is not moving, the machine will stitch in the same place many times and make a thread knot that the needle can not penetrate. This will break the needle. Is there adequate presser foot pressure to hold the fabric in contact with the feed dogs? You may need to adjust the presser foot pressure if your machine has adjustable presser foot pressure. If the fabric is not moving through the machine try running the machine with no thread and see if the fabric will move through the machine. Sometimes tangled thread can stop the fabric from moving through the machine.

Check that the feed dogs are moving. To do this first un-thread the machine and turn the handwheel so that the needle is in the raised position. The feed dogs should also be in the raised position like in the first picture below.

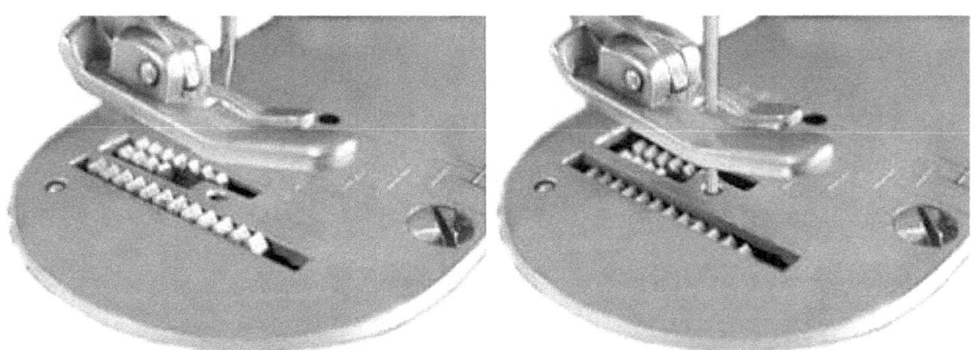

As you continue to turn the handwheel the feed dogs should move to the back of the machine and then as the needle moves down the feed dogs should also move down below the needle plate as in the second picture. The feed dogs will then move toward the front of the machine before raising to start the next cycle

If the feed dogs are always in the lower position, check to see if your machine has dropping feed dogs. There may be a feed dog dropping knob and it may be in the down position. Check your owners manual, some machines have the feed dog dropping knob in a strange location like under the machine or in the back of the machine. If your machine does not have a feed dog dropping knob or if it does, but the knob is not in the dropped position and the feed dogs are still too low to move the fabric then your machine has a problem and needs to go to the shop for adjustment or repair. If the feed dogs are not moving at all your machine also needs repair.

Some seams are too thick for the feed dogs to handle. If you are trying to sew a thick seam and the feed dogs are not able to move the fabric over the seam there is a way to manually advance the machine, see the section "Tips and ideas" in the chapter "How to use a sewing machine". Make sure that you are not trying to sew seams that are too thick for the type of machine you are using. You may need to use a heavier machine.

6. Needle centering - As the needle moves up and down the needle should not come too close to the edge of the hole in the needle plate or the toes of the presser foot. If the needle touches or hits the needle hole or pressure foot it is bad news, the needle may bend or break. At the very least you will get thread breakage and skipped stitches. To check the needle centering first un-thread the machine. Rotate the handwheel of the machine a few times and watch the needle move up and down through the needle hole in the needle plate. Is the needle coming too close to the edge of the hole in the needle plate or is the needle actually hitting the needle plate? If it is then you may have a bent needle, try replacing the needle with a new one. On a zigzag machine (like in the picture below) the needle hole is oval shaped and the needle may move from side to side, this is normal as long as the needle does not come too close to the edge of the needle hole. On some home machines the needle clamp can have some play (some movement) that allows the needle to move slightly from side to side. Try loosening the needle clamp and moving the needle from side to side so that it is correctly positioned in the center of the needle hole and then re-tighten the needle clamp. Some decentering is OK, we just don't want the needle so close to the edge of the needle hole that any slight bending will cause the needle to strike the needle plate and bend or break. If the needle is still not centered then your machine will need to go to the shop for repair.

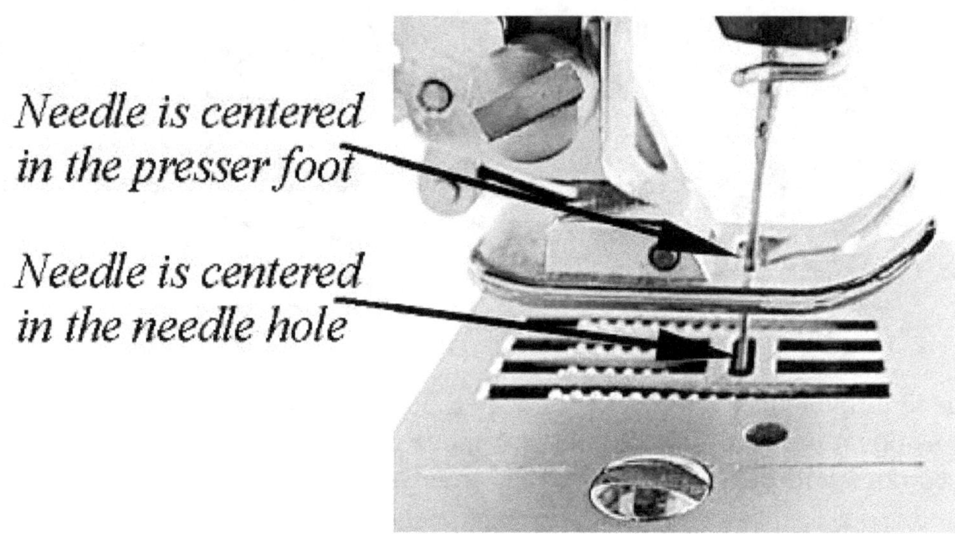

Could not find the problem? If none of the above checks isolate the problem then you may want to take your machine to a service shop. If you want to try to repair the machine yourself then get our other book, "The Sewing Machine Master Guide".

Fabric bunching

1. The stitch length may be too long. To keep thinner fabric from bunching a shorter stitch length is needed.
2. The thread tension may be set too tight. Try to loosen the thread tension slightly.
3. If you are using a zigzag stitch try a triple stitch if your machine has that stitch.

Loose stitching

1. Check for correct thread tension and bobbin tension. See the chapter Adjusting Tension.
2. Go through the "Needle thread breaks or stitch problems" section.
3. Go through the "Bobbin thread breaks or stitch problems" section.

Machine runs slow or is noisy

1. Oil the machine (unless you have an oil free machine, see your users manual).
2. Clean the hook-race and feed dog area.
3. Check for tangled thread in the mechanical parts of the machine.
4. Go through the Machine Checklist in the section "In-depth troubleshooting" later in this chapter.

Machine does not run

1. If the motor runs but the hand wheel does not turn then the drive belt is broken or loose. See the section "Drive belts" in the chapter "Maintenance".
2. If the motor makes a humming or buzzing sound but does not run then the motor or the machine is frozen or jammed. You may want to take your machine to a service shop. If you want to try to repair the machine yourself then get our other book, "The Sewing Machine Master Guide".
3. If the machine makes no sound or has no power then make sure that the foot pedal is plugged in to the machine. Check to make sure that the machine is getting power (is plugged in to a working wall socket). If the machine is still silent take the machine to a service shop.

Maintenance

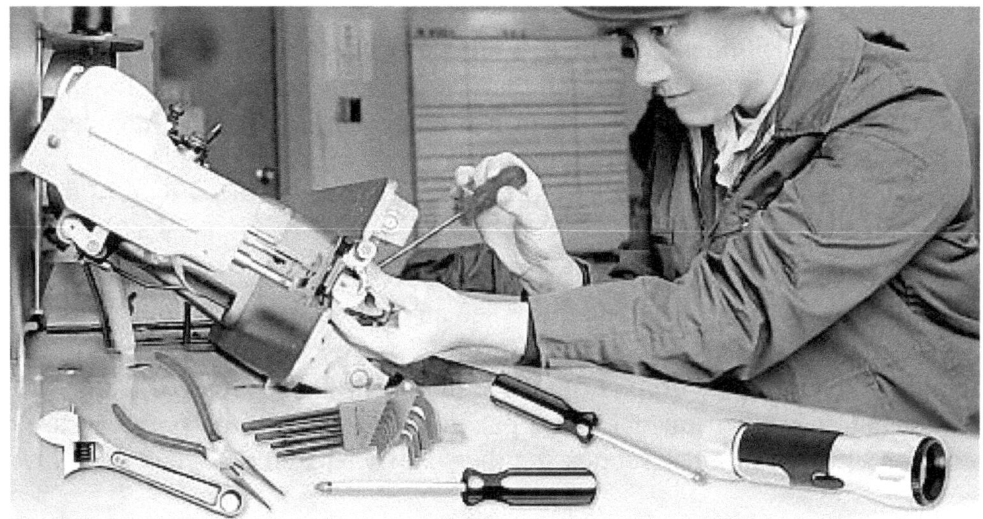

This chapter covers the basics of sewing machine maintenance. The procedures in this chapter may not exactly apply to your machine, some electronic machines are substantially different then the generic machines depicted here. Always follow the instructions in the owners manual for your specific machine and if any part of the procedures presented in this chapter do not seem appropriate for your machine you should give preference to the procedures in the owners manual.

Maintenance intervals

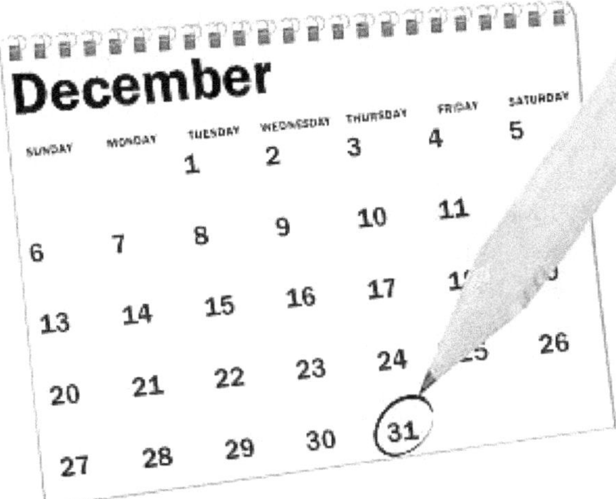

Every 8 to 12 hours of actual running time or every 2 years do the following:

- Clean and oil
- Replace the needle
- Run the machine and listen for problems
- Test sew

Lubed for life?

Many newer sewing machines made within the last 20 years are so called "lubricated for life". If you have one of these machines the manufacturer will state in the owners manual that the machine does not require lubrication or that only one or two places such as the hook race and needle bar require lubrication.

- Actually there is no such thing as lubricated for life, what they really mean is that the machine is intended to have a limited service life and that the lubrication should last for the intended service life. However, some of these machines can last much longer then the intended service life. For average home use "Lubed for life" really means "Lubed for 5 to 15 years"
- The lubrication that is used in these machines is a synthetic or other very high quality lubricant and actually does last a very long time.
- My advice concerning "lubed for life" machines is to inspect the bearing areas that you can see (like the needle bar area) and see if you can see any lubricant around the bearings. If so, does it look reasonably fresh or does it look dried or dirty? If you can see lubricant that looks OK then you can do nothing or put in a drop of oil. If the lubricant that you see looks dry, dirty or you can't see any lubricant at all then oil normally, put in enough oil so that you can see that it is well lubricated. For pictures and instructions see the section later in this chapter "How to oil moving parts".
- Do not over oil "lubricated for life" machines because you do not want to flush out all of the original lubricant, most of the time the original lubricant starts to get thick and a drop or two of oil will thin it out and restore proper lubrication.

Oil for sewing machines

Caution - Use only sewing machine oil in your sewing machine. Do not use "3 in 1 Oil", WD40 or any other type of household oil in a sewing machine. Household oil will become thick and gum up your sewing machine. Sewing machine oil is inexpensive and is available at any sewing machine store or even at Walmart in the sewing section of the store. If you have put household oil in your sewing machine then just re-oil the machine with sewing machine oil to fix the problem, make sure to wipe up any extra oil that drips out!

Oil Bottles - Get one of the small 4 ounce plastic bottles with the long pull-out tube like the one pictured below if you can, it makes oiling much easier. The bottles with a pull out tube are more expensive but it is worth the cost to get the pull out tube. With the long tube you can put the oil exactly where it is needed in the right amount, otherwise it is hard to get the oil into the right place and you end up dripping a lot of extra oil into the machine and all over your table and floor.

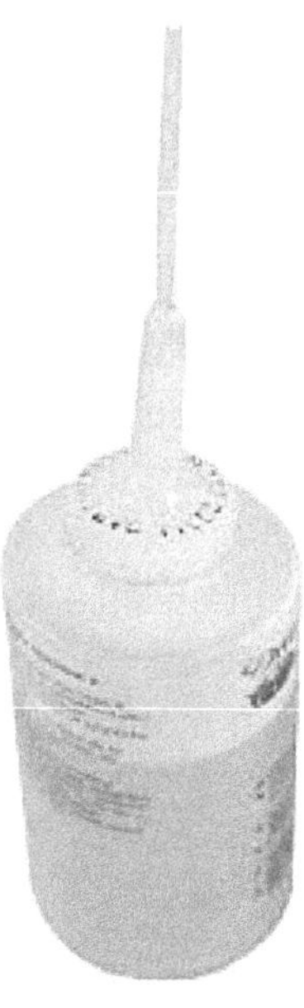

How to oil moving parts

Basic oiling procedure - The picture below is an example of some metal to metal and plastic to metal bearing surfaces that require oil. If you look closely at the picture you will see that some of the bearings have no visible oil around their outer edges (the part you can see). For these dry looking bearings you should start with 2 drops of oil and let the oil seep into the bearing for a few seconds. If the bearing still looks dry then add another drop until you can see a slight hint of oil around the outer edge of the bearing. The bearings in the picture labeled as "Oil is visible" have some oil that you already can see. For these bearings you need to determine if the oil looks old and thick or looks more like new oil. If the oil looks old and thick you should add a drop or two. If the oil looks thin and new you can pass over that bearing, no oil is needed. The bearings in the picture had oil that was kind of thick so I added a drop or two of oil.

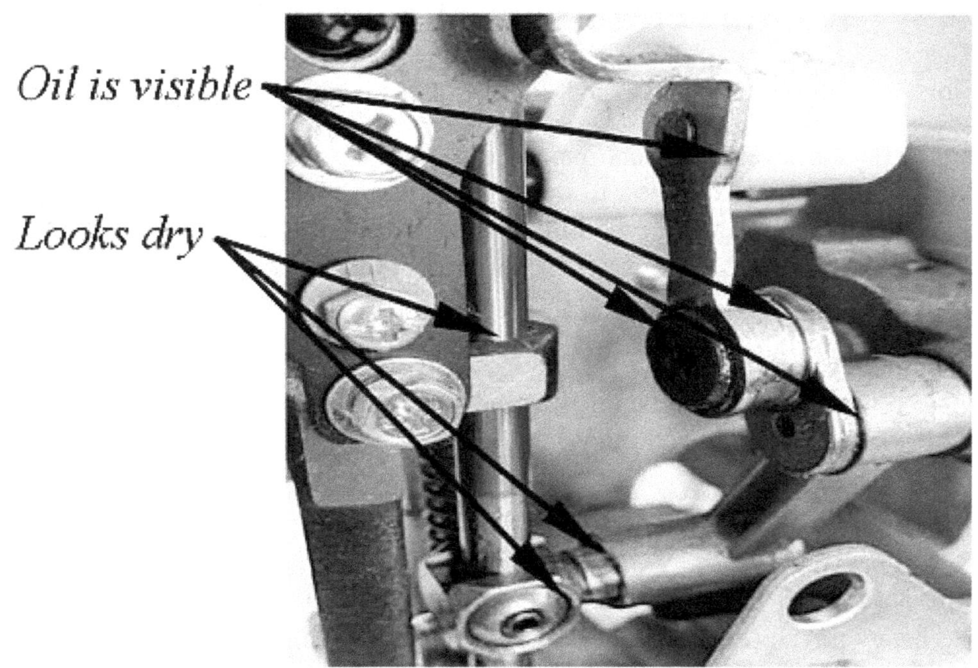

- Put a drop or two of oil in the areas that your owners manual tells you to oil. If you have an older machine and do not have an owners manual then oil all bearing surfaces that you can see (see pictures below for an example). Bearing surfaces are any place that moves when the machine is running and has metal to metal contact such as the needle bar, linkages, shafts, etc. Some moving parts in modern machines are made from a kind of hard nylon (a type of plastic). These parts typically are lubricated the same as metal parts.

- If you see gears that have grease, you should not oil them or if the grease looks dry you may try using one or two drops of oil to thin the grease.
- Over oiling will cause oil to drip from your machine and make a mess, but otherwise will not harm your machine unless you get oil into the motor. On most bearings if you put in a few drops of oil you will see a little oil around the edges of the bearing surface and that tells you that it has enough oil. If the bearing is really dry it will take in more oil, if the bearing already has oil then just a drop or two will be enough.
- On older machines make sure to oil the hook race (see the pictures in the next section). On newer machines with a plastic bobbin-case do not oil the hook race unless the owners manual specifically tells you to oil it, most plastic bobbin-case machines do not take oil in the hook race.

Belts and electronics - Do not get oil on belts or drip it into electronic parts or onto circuit boards. Clean up excess oil with a cloth and q-tips.

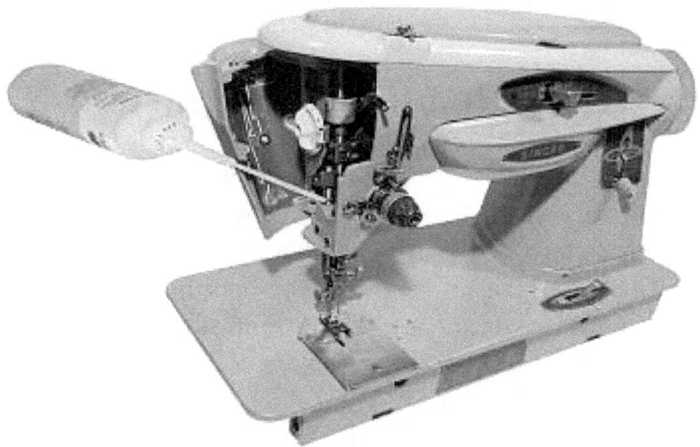

Routine clean and lube - step by step

Don't forget to look over the section "Oil Points" a little later in this chapter.

Owners manuals - If you have an owners manual available for your machine then follow the cleaning and oiling instructions it gives in addition to the instructions in this section. Every machine is different and the owners manual may give you needed information for your specific model of machine such as if your machine has an oil free hook race or not. Also some newer machines do not require periodic oiling or only require oiling a few points, the owners manual will tell you this.

NOTE: Always unplug the machine before doing any work, this includes cleaning and oiling! If you need to run the machine for a minute to let the oil flow into the bearings then plug it in for that, but unplug it again to continue with your cleaning and oiling.

Upper thread path

- Follow the Thread path and clean lint and old thread from all thread guides.
- Get some heavy thread (button thread works well for this) and run it through the tension disks to clean lint and old thread from the tension disks. Do this as you would use dental floss to clean your teeth. Be careful not to damage the delicate take-up spring while you clean the tension disks.

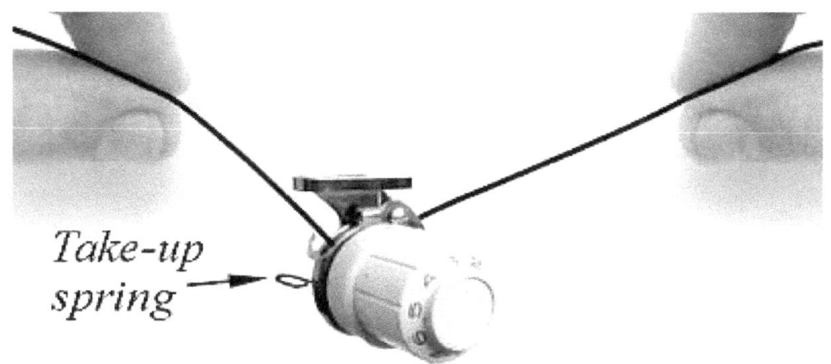

Needle bar area

- Remove the needle bar cover. On some machines the needle bar cover just swings open for easy access (like the machine pictured below), but on other machines you will have to remove a screw or a few screws.
- After the needle bar cover is removed use a brush and q-tips to clean lint, thread and dirt from the needle bar and take-up lever areas. If you see clean looking oil or grease on any of the parts, you don't need to wipe it off.
- Oil all bearing surfaces that look dry or in need of oil. If you are not sure, try adding a drop of oil at a time until you see some slight amount of oil around the edges of the bearing surface to let you know that the bearing is no longer dry.
- During routine cleaning and lubrication it is a good idea to replace the needle.

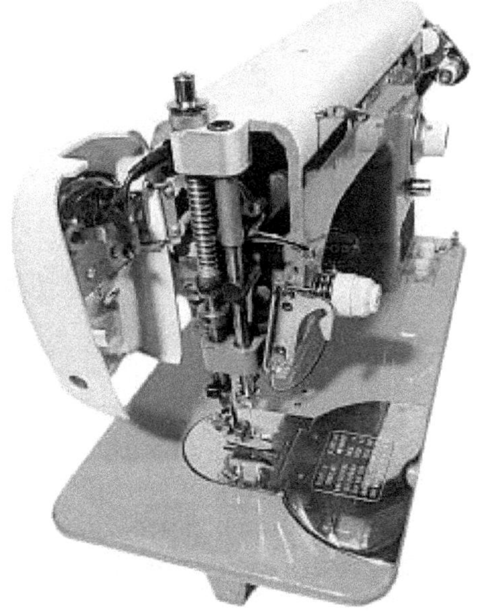

On the machine pictured below one screw was removed and another screw was loosened to remove the cover.

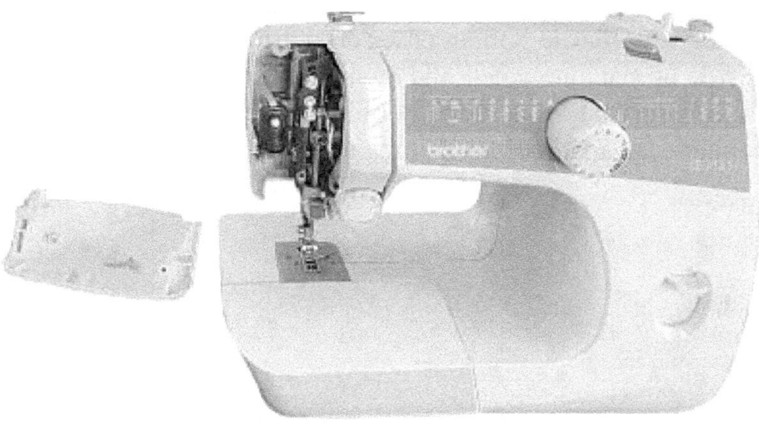

If your machine has a 2 part plastic case you do not need to remove the entire case for routine cleaning and lubrication. The machine pictured above has a 2 part plastic case. It is called a 2 part plastic case because the main body of the case comprises of a front half and a back half (kind of like a clam shell). There are also removable covers over the needle bar area and hook area so that routine maintenance can be done.

If you have an older machine that has easy access to the bottom parts of the machine then you should clean and oil all of the parts that are accessible during routine maintenance. Some older machines also have side, top and back panels that remove with one or two screws, in this case also clean and oil moving parts that are accessible through these panels.

Hook and feed dog area

For machines with top loading bobbins:

- Remove the needle plate (and hook cover plate if it is a separate plate) if you are able to do so. The needle plate is removable on all types of machines including machines with a 2 part plastic case. On most machines you will need to remove a few screws to take off the needle plate but some modern machines have needle plates that snap out or lift off. The machine pictured below has two large screws that must be removed to take off the needle plate.

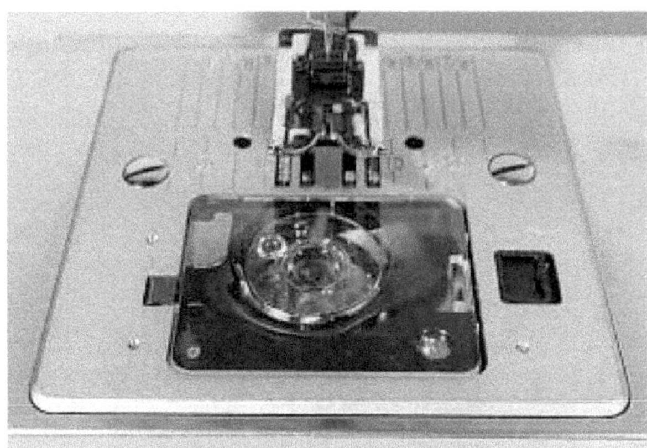

- After the needle plate has been removed use an old tooth brush and Q-tips to clean lint and dirt from the feed dogs and feed dog area. You can also use a cleaning brush (they look like a small paint brush with stiff bristles). Remove the bobbin case if it is removable on your machine. Clean the hook area and bobbin case area. A vacuum cleaner will help if one is handy.
- Oil all bearing surfaces that look dry or are in need of oil.

The next picture shows same machine as pictured above with the needle plate, bobbin cover and presser foot removed. This machine has a plastic bobbin case. The plastic bobbin case is the large round black plastic part that the bobbin goes into. The hook race is the large circular metal ring that is around the bobbin-case. Most machines with plastic bobbin cases do not require oil in the hook race, you should not oil the hook race on this type of machine unless the owners manual specifically tells you to oil it. If on the other hand your machine is older and has a metal bobbin case then you do want to put a few drops of oil at the point where the bobbin-case and hook race come together.

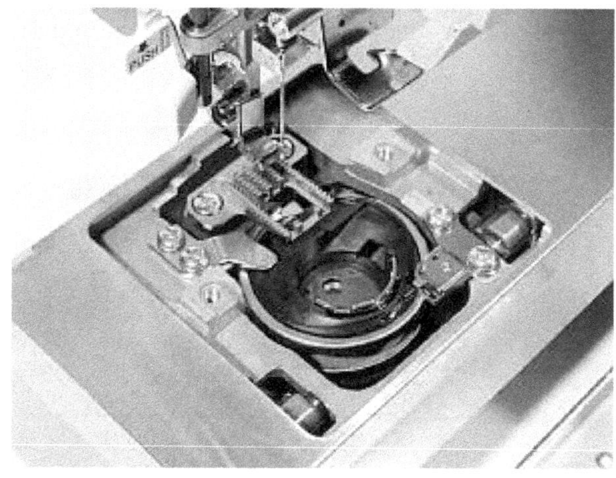

For machines with front or side loading bobbins:

Machines with an HA-1 type bobbin case require oil in the hook race. The first picture below shows an assembled HA-1 type hook. The race cover is removed by rotating the two black levers to the outside and simply lifting the race cover out. Once the race cover is removed the shuttle-hook can also be simply lifted out. Remember how the shuttle-hook was seated so that you can put it back togather!

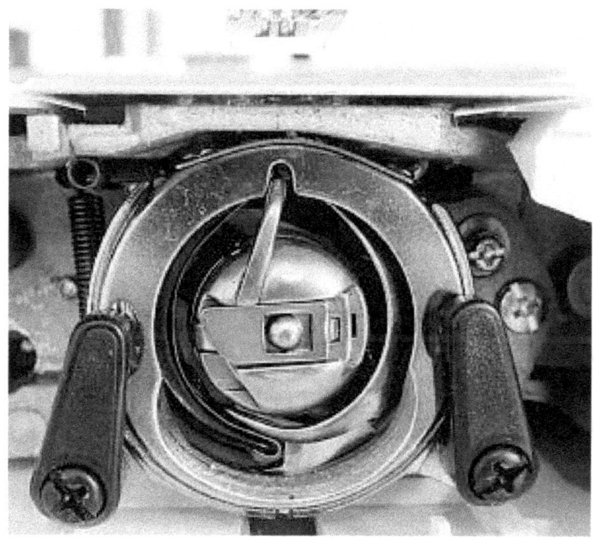

The next picture shows the same machine with the shuttle-hook and bobbin case removed.

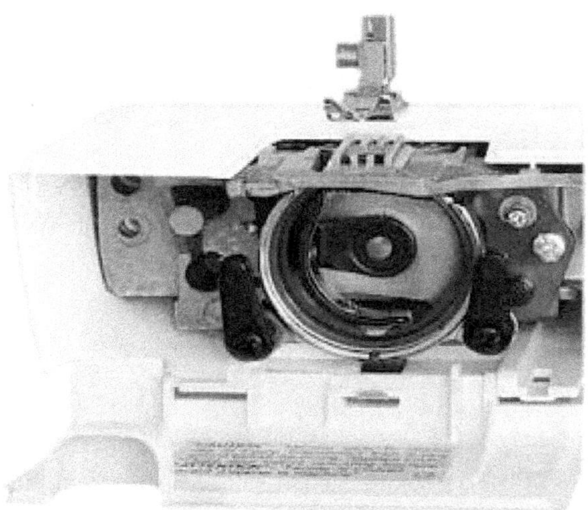

Once shuttle-hook is out of the machine then the race and shuttle-hook can be cleaned with a cloth and a few drops of oil can be placed on the shuttle-hook (shown in the picture below). Not that this picture is showing the back of the shuttle-hook, this is the side that goes into the machine first.

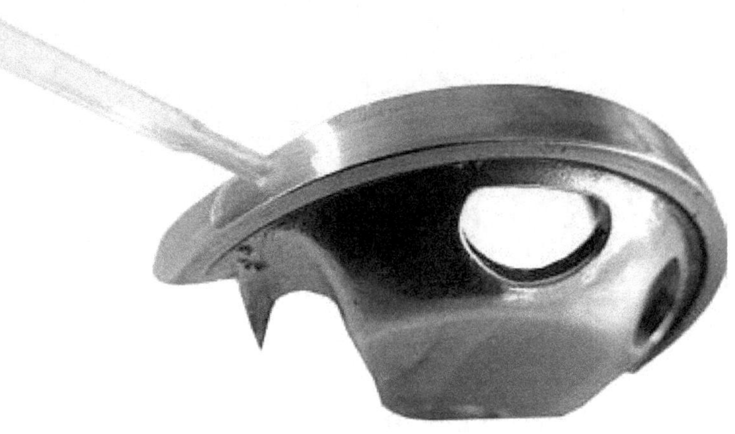

Oil Points

The picture below shows the oil points in the needle bar area of a typical zigzag machine, not all oil points (moving surfaces with metal to metal contact) are shown in the pictures because some are not visible. You should look from all angles as you rotate the handwheel of the machine to find bearing surfaces that require oiling.

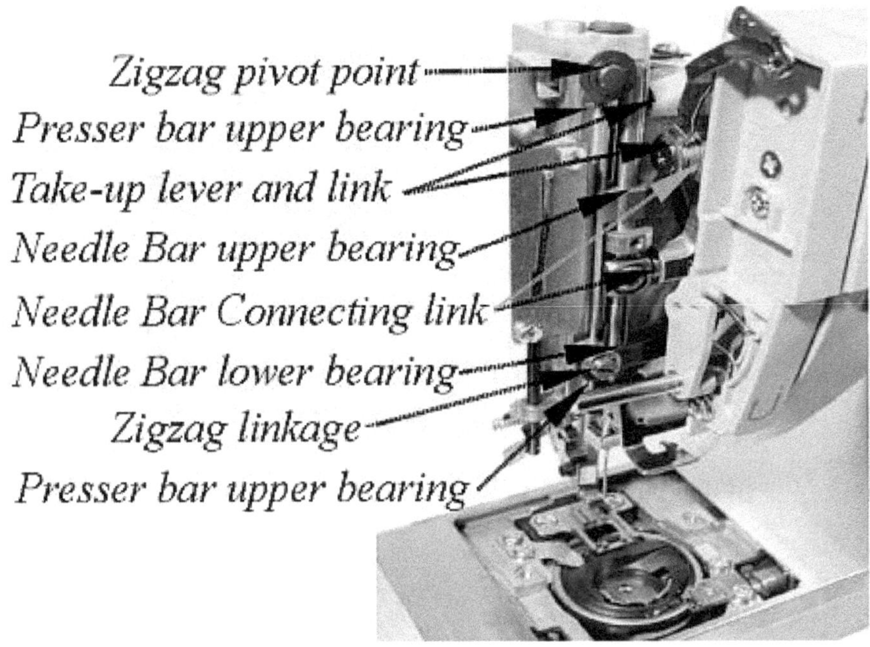

The next picture shows the oil points of a common HA-1 type machine. Note that the machine pictured in the diagram has a side facing hook but the machine in the pictures above has a front facing hook. The oil points are basically the same.

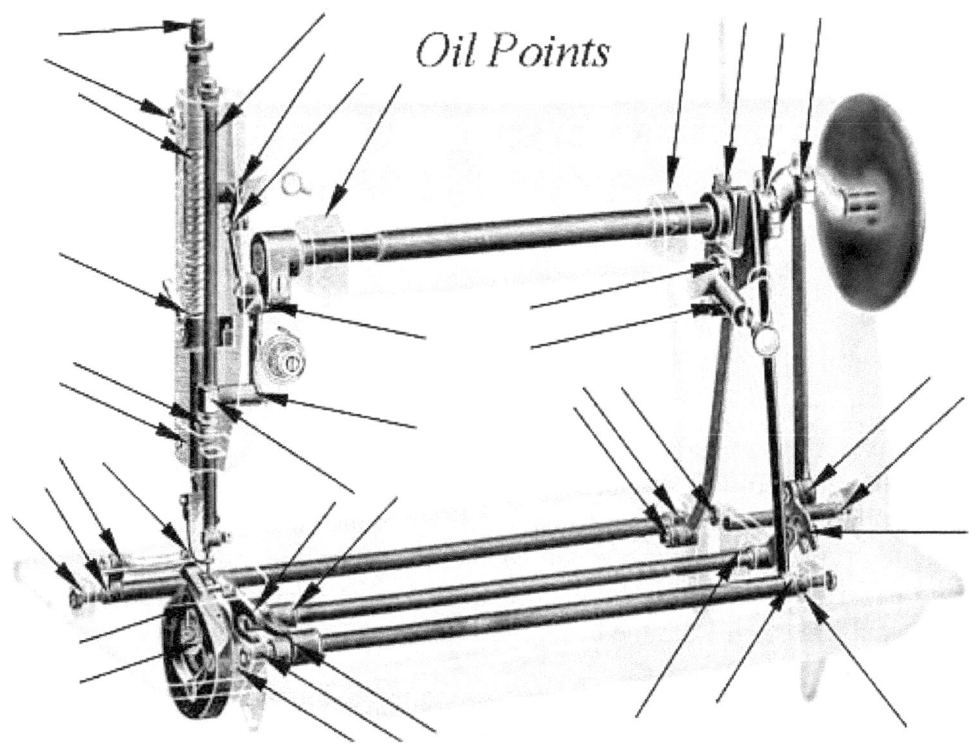

Drive Belts

Drive belts come in three general types; round belts, v-belts and cogged belts.

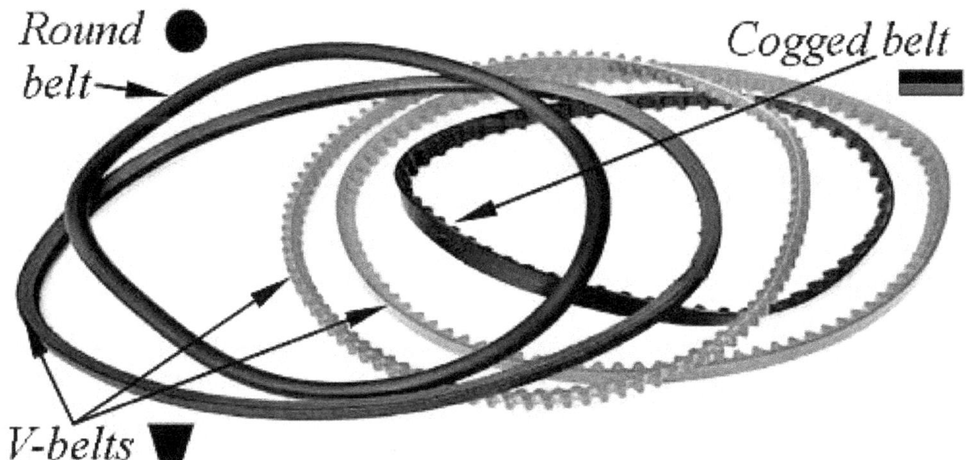

Round belts have a round cross profile (if you cut the belt crosswise and look at the cut end you are looking at the profile). Round belts are made of rubber and are generally used on older machines. Round belts are the least efficient at transferring power, so after about 1950 manufacturers started using V-belts.

V-belts have a modified "V" profile with angled sides and a flat bottom. They are used on home machines from about 1950-1990. The two lighter colored V-belts shown in the picture below have undercutting to make them more flexible, it may look like they have teeth or cogs but these belts have smooth angled sides and really are V-belts.

Cogged belts have teeth (also called cogs) and require a pulley that has corresponding teeth. They are used on most machines built in the 1990's and later. Cogged belts do not slip because the teeth of the belt interlock with the teeth of the pulley. They are made with a rectangular profile and are thinner and more flexible than V-belts and therefore more efficient at transferring power. The picture below shows a cogged belt and cogged pulley in a modern machine.

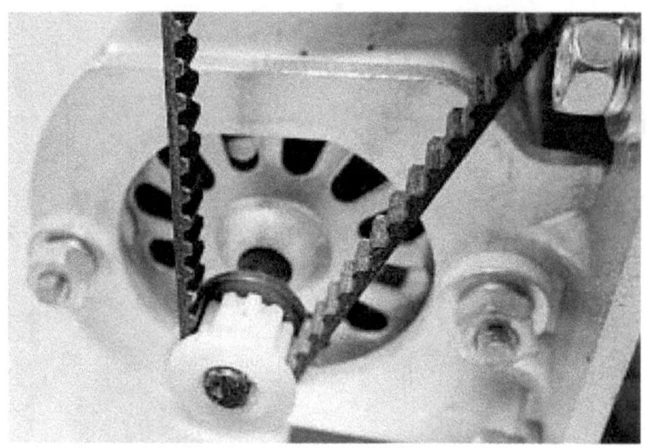

Belt Inspection - Look at the belt and see if it is in acceptable condition. If the belt is coming apart or is chewed up it should be replaced. For a V-belt he surface that you should be most concerned with is the surface of the sides of the belt that comes in contact with the pulleys. If the surface is shiny (this is called glazed) then the belt will slip. In this case the belt should be replaced or you can try to lightly sand it with some 220 grit sand paper to remove the shiny surface. If you do this make sure to clean the belt with a cloth to remove the sand paper grit from the belt before trying it on the machine.

Tension adjustment - On most machines the belt is adjusted by tilting or moving the motor on its mounting system. There is usually an adjustment screw or screws to do this. On some machines there are several mounting nuts that must be loosened to adjust the belt. Some older machines have a spring loaded motor mount that does not require adjustment and keeps to belt tension at the correct level automatically.

Belt tightness for V-belts and round belts - If your machine has a smooth drive pulley on the motor (most older machines do) then your machine uses a v-belt or a round belt. These belts should be only tight enough to prevent slippage. If belts are too tight they rob power and slow the machine. Also over tight belts cause unnecessary bearing wear and premature belt failure. Belts that are too loose slip, cause poor acceleration of the machine and cause the belt to prematurely wear out. Start with the belt too loose and then tighten it until the belt no longer slips under acceleration. The belt should be just tight enough to avoid belt slippage but not tighter. You can test for belt slippage by holding the handwheel and pressing on the foot pedal to run the motor for a second or two. This is called a stall test. In this case a little belt slippage is normal in most cases. While you are sewing, if you notice belt slippage then adjust the belt a little bit tighter until you can sew without belt slippage.

Belt tightness for cogged belts - If your machine has a pulley with teeth and your belt also has teeth then you have a cogged belt. This type of belt is used on most new machines including all electronic machines. Adjust a cogged belt so that it is not loose but not overly tight either. Real scientific huh? Actually the correct tension depends on several factors including the distance between the pulley and the handwheel and the design of the belt. You want the belt to have enough tension so that the teeth of the motor pulley stay connected with the teeth of the belt and prevent the belt from slipping. If there is too much tension it will cause rapid wear of the motor bearings and the belt. So it is a compromise. If you are not sure you could try to hold the motor pulley from moving and try to turn the hand wheel and see if you can cause the belt to slip. If you can't cause the belt to slip on the motor pulley then the belt is probably tight enough. Make sure that the belt is not too tight, if you pluck the belt like a guitar string it should vibrate at a low frequency like a bass guitar or a really low humming sound. If it is too tight it will vibrate at a higher frequency like a high humming sound.

Bobbin winders

If the winder tire has a flat spot or other problem then the bobbin will not wind smoothly and the tire must be replaced. The next picture shows a bobbin winder on a newer style machine with the case removed. During operation the winder swings to the right so that the tire is held against the contact surface of the handwheel pulley. On this machine there are no adjustments provided, so if the winder does not work correctly it should be replaced.

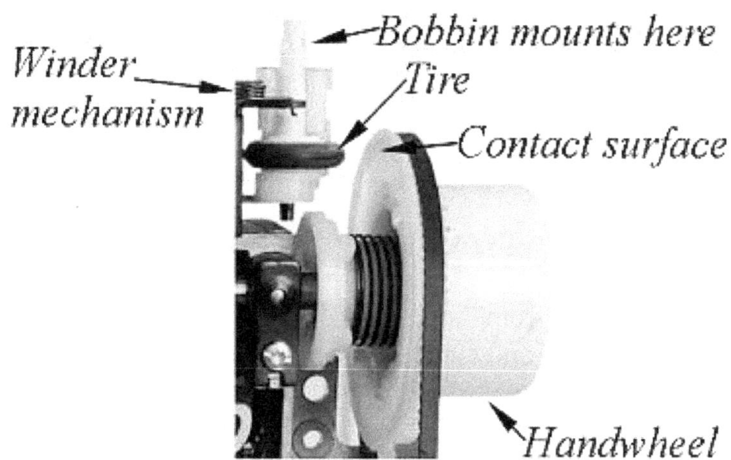

Spin the bobbin winder shaft with the winder disengaged (non-winding position). On most machines the bobbin shaft will spin smoothly and freely. Oil if needed. Remove any excess oil so that it does not get on the bobbin winder tire and cause it to slip. On some machines the bobbin winder will not spin freely when it is disengaged (there is a brake that is engaged in the stop position). If this is the case then just oil the bobbin winder shaft and make sure that the winder works smoothly when engaged.

Lights

LED Lights - Some newer machines use LED lights. These lights very rarely go bad, if your machine has LED lights and they don't work then it is probably a power supply issue and beyond the scope of this book, someone with experience troubleshooting electronics should troubleshoot it.

Incandescent Lights - Most older machines and some newer machines use incandescent lights. These are small light bulbs that look similar to a night light bulb. Make sure that you replace the bulb with the correct wattage bulb. If you put in a bulb that is over the specification for your machine then the light will overheat and could melt the wiring or other parts or could be hot enough to burn you. When you remove the old bulb you can usually find the wattage written on the bulb or the base of the bulb. Some machine have the recommended wattage on the name plate of the machine or on a label in the area of the light. Most machines take a 15 watt bulb. There are two types, screw-in base and bayonet style base.

- The screw-in type is just like a miniature normal light bulb and unscrews in the counterclockwise direction. If the bulb does not unscrew and feels stuck then you may have a bayonet type.
- The bayonet type bulb must be pushed in against a spring and then rotated in the counterclockwise direction about 1/8Th turn to remove (after 1/4Th turn it will just pop out). To replace you must align the pins of the base with the slot in the socket and then push the bulb in against the spring, then turn clockwise about 1/8 turn to lock in the bulb. A bayonet type bulb is pictured below, on the right it has been inserted into the socket but has not yet been turned clockwise 1/8th turn to lock it in.

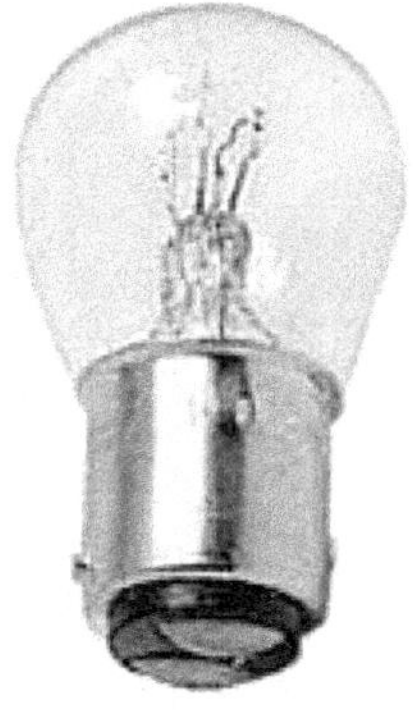 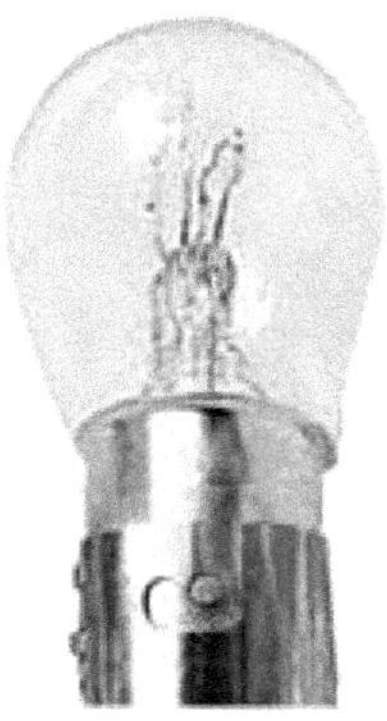

Links & Information Sources

The following are various sewing machine related links you may find helpful.

New machines

I buy most of my new home machines from Amazon and have not found another seller that I consistently use. You should check out other sellers and compare for your self though, you may find other sellers that work good for you.

amazon.com

The top selling sewing machines on Amazon

amazon.com/gp/bestsellers/sewing-machines

The top selling sergers on Amazon

http://www.amazon.com/gp/bestsellers/sergers

Used machines

http://www.ebay.com/

http://www.craigslist.org/about/sites/

Home machine parts

http://www.sewingpartsonline.com/

http://www.sew4less.com/index.php

http://sewcat.com/

http://www.autumnsewing.com/

Home sewing machine manufactures

babylock.com

http://www.bernina.com/

http://www.brother-usa.com/homesewing/

http://www.elnausa.com/

http://janome.com/

http://www.jukihome.com/

http://www.pfaff.com/

http://www.singerco.com/

http://new.husqvarnaviking.com/

Sewing machine history

http://www.ismacs.net/home.html

http://www.ismacs.net/singer_sewing_machine_company/model-list/index.html

http://www.sewalot.com/sewalot_index.htm

http://urbanglasgow.co.uk/archive/clydebank.-past-present.__o_t__t_803.html

http://www.singersewinginfo.co.uk/index/

http://www.moah.org/exhibits/virtual/sewing.html

Sewing machine Theory and Information

http://en.wikipedia.org/wiki/Category:Sewing_machines

http://en.wikipedia.org/wiki/Sewing_machine

http://www.juki.co.jp/industrial_e/customer_e/books_e/index.html

http://www.fashion-incubator.com/

http://www.uspto.gov/web/patents/classification/uspc112/defs112.htm

Sewing machine manuals

http://www.ismacs.net/free-sewing-machine-manuals-list.html

http://parts.singerco.com/html/inst_manuals.html

http://www.singerco.com/support/replacement-parts-chart

http://parts.singerco.com/IPinstManuals/

http://parts.singerco.com/IPsvcManuals/

http://www.dixiesewingmachine.tzo.com/SERVICE%20MANUAL.htm

http://www.sewingmanuals.com/index.htm

http://www.sewusa.com/Sewing_Machine_Instruction_Manuals.htm

http://sewingwishlist.com/janome.html

http://riccar.com/help/manuals/

Needles

http://organ-needles.com/english/

http://www.beisselneedles.com/index.html

http://www.schmetzneedles.com/

Embroidery

http://www.embird.net/

http://www.open-source-embroidery.org.uk/osembroidery.htm

http://sourceforge.net/apps/mediawiki/kxstitch/index.php?title=Main_Page

http://www.melco.com/

Quilting

http://www.gammill.net/index.php

http://www.handiquilter.com/

http://quiltfrog.com/

http://www.nolting.com/index.php

http://www.quiltingboard.com/forum.php

http://en.wikipedia.org/wiki/Longarm_Quilting

Thread

http://www.amefird.com/

https://www.guetermann.com/shop/en/view/content/

http://www.coatsindustrial.com/en

Glossary

A

Accessories - Most accessories are types of feet that mount on the presser bar of the sewing machine in place of the standard straight stitch or zigzag foot. Accessories allow the machine to do something that it can not ordinarily do or make it easier to do something that is ordinarily difficult. Some examples of accessories include buttonholers, zigzagers, piping feet, edge cutting feet, ruffler feet, etc.

Accessory Mounting Point - A threaded hole (or holes) on the bed of the machine to mount plates or guides. Some examples of accessories that mount in this manner are edge guides and feed dog cover plates.

Alpha Numeric Characters - Digital numbers and letters. Some machines are capable of sewing decorative stitches that incorporate numbers into the stitch. This can be used for decoration or for monograming.

American & Efird - A large manufacturer of high quality thread. Produces many types of threads including advanced and technical threads.

Automatic Back Tacking (or automatic tacking) - This is the automatic sewing of reinforcement at the beginning or end of seams to prevent the stitches from pulling out. See Reinforcement.

Automatic Needle Threader - A device to aid in the threading of the sewing machine needle. Most are not completely automatic but only aid in threading the needle, the thread is placed and held over the threader and a lever is pushed to cause the threader to move into place over the needle and make the thread go through the eye of the needle. The threader is then retracted.

Automatic Needle Up/down - This is a function that is available on electronic machines. There is a button that when pressed will cause the needle bar to move the needle to the down position or the up position. In addition most electronic machines can be set to always stop with the needle up or down. For making pivot turns (using the needle as a pivot point) it is handy to set the machine to stop with the needle in the down position. For regular sewing the machine would be set to stop with the needle in the up position.

Automatic Presser Foot Lifter - This function is available on some more expensive home machines and on some industrial machines. The machine will automatically lower the presser foot at the start of the seam and raise the presser foot at the end of the seam. Most machines that have automatic presser foot lift capability can be programmed to raise the presser foot and stop with the needle in the down position for making pivot turns. This function can also be combined with automatic back tacking and/or automatic thread cutting on some machines.

Automatic Reinforcement - See Automatic back tacking.

Automatic Thread Cutter or Thread Trimmer - This function is available on some electronic home machines and on some industrial machines. The thread is cut automatically at the end of the seam, when the machine is stopped or when a button is pushed (depending on the capability of the machine and how the machine is set to operate).

Auto-sizing Buttonholer - An automatic buttonholer function of a sewing machine that uses a special foot to detect the size of the button and make the correct size button hole for that size of button. Typically the button is inserted into the special foot and then the button hole is made.

B

Balanced Stitch - In the case of a lockstitch, the stitch is balanced when the lockpoint is in the center of the fabric. The lockpoint is the place in the stitch where the needle thread and bobbin thread cross. If the stitch is unbalanced the needle thread or bottom thread is too tight or loose. If the needle thread is too tight it will pull the lockpoint and the bottom thread up to the top of the fabric. If the bottom thread is too tight it will pull the lock point and needle thread down to the bottom of the fabric.

Bar Tack - A bar tack is a length of closely spaced zigzag stitches (usually about 10mm to 20mm long) that is used to reinforce areas of stress or weakness in garments or sewn products. A bar tack is used to reinforce the openings for pockets on jeans, to attach belt loops and straps, secure the ends of webbing, etc.

Basting - Also called tacking, this is a temporary seam used to hold fabric in place until permanent construction seam can be used. The basting stitch is made in such a way that it can easily be removed, with a lockstitch sewing machine this can be accomplished by loosening the needle thread so that the lock will be pulled to the bottom of the fabric. This allows the bottom thread to be pulled out later. This is usually done when assembling a complicated design so that the parts can be roughly tacked together and once they are held in place the permanent stitching can be done without the parts moving around. Sometimes a basting stitch is used for the cuffs of pants because this allows the stitch to easily be removed if the hem needs to be raised of lowered. Industrial basting machines are available that use a single thread chain stitch that can be easily removed (stitch type 101).

Basting Stitch - See Basting.

Bed - This is lower flat or cylindrical part of the sewing machine that the fabric travels across during sewing. Flat bed machines use a flat bed and are used to sew fabric that is laid flat during sewing. Cylinder bed or post bed machines use a smaller cylinder shaped bed that can accommodate sewing tubular shaped fabric such as sleeves, pants legs or shoes.

Belt - see drive belt

Bearing - A bearing is mechanical device used support a part and allow movement of that part such as a shaft that rotates. If the shaft were just bolted to the machine it could not move or rotate. Instead a bearing is attached to the machine and then the shaft is inserted into the bearing allowing it to rotate in a precise position. In a sewing machine most bearings are rotational but some are sliding such as the needle bar and presser bar. There are many types of bearings such as the sleeve bearing, ball bearing, roller bearing and cone bearing. Most bearings use some form of lubricant (oil or grease). The lubricant goes in between the metal or plastic surfaces of the bearing to prevent the surfaces from actually touching. In this way wear of the load bearing surfaces of the bearing is reduced. There are open bearings and sealed bearings. Open bearings are open to the air. Sealed bearings have a seal that is usually made of rubber to close the bearing from the open air and to contain the lubricant. This greatly prolongs the life of the lubricant but makes it more difficult to add lubricant later.

Bernina International - A manufacturer of home sewing machines and embroidery machines based in Switzerland.

Blind Hem - A hem made with a blind stitch so that the stitching is not visible from the outside of the garment. Blind hems are also used in draperies.

Blind Stitch - The blind stitch is used to do hemming of pants and dresses and is used because it is less visible then regular types of stitches. There are several variations of the blind stitch depending on what kind of machine is used to make it. 1. If made by a zigzag machine then this it is a variation of the zigzag stitch in which a straight stitch is made for several stitches and then a single zigzag stitch is made followed by several more straight stitches and so on. 2. If made by a blind stitch machine it is a modified zigzag stitch made using a curved needle that only penetrates the back of the fabric leaving no visible stitches on the front of the fabric. 3. If made by hand it is similar to #2.

Bobbin - A small spool that is comprised of a cylinder with flanges on which thread is wound. A bobbin is used in lockstitch sewing machines for the bottom thread instead of a larger spool because the bobbin must pass through the loop during the formation of each stitch and a larger spool would be too large to pass through the loop.

Bobbin Case - A bobbin case is the part of a sewing machine that holds and retains the bobbin.

Bobbins Prewound - Prewound bobbins are wound with thread from the factory. They are available for both home and industrial machines. They are wound using precision high speed winding machines and typically hold up to 25% more thread then a bobbin that is wound on the sewing machine. Most prewound bobbins are wound on disposable paper bobbins with very thin flanges or no flanges. Prewound bobbins are not overly expensive if purchased in large quantities. The total cost of using them may actually be less then standard bobbins because the machine does not need to stop as often to change bobbins and the operator does not need to spend time winding bobbins.

Bobbin Thread - The bottom thread on a lockstitch sewing machine.

Bobbin Thread Warning - Some electronic machines have a sensor to alert you when to bobbin thread is getting low or has run out.

Bobbin Tension Spring - The spring that controls the bobbin thread tension, it is usually a small screw on the bobbin case.

Bobbin Type - Bobbins come in various sizes and these sizes have type letters or type numbers to identify them (also called class numbers). For example industrial machines use type L, type M and many others. Home machines use class 15, Class 66 and many others. You must use the correct bobbin type or class for your machine, they are not interchangeable. Some home machines also use the industrial type L bobbin.

Bobbin Winder - A device that is built into a sewing machine to wind bobbins.

Bobbin Winder Tire - A small rubber tire mounted on the bobbin winder shaft of a sewing machine that contacts the rotating hand wheel to turn the bobbin winder during bobbin winder operation.

Bobbin Winding Clutch - A device to disconnect the handwheel of a sewing machine from the main drive shaft during bobbin winding so that the parts of the machine will not run, only the motor and handwheel will turn. This cuts down vibration and wear on the machine during bobbin winding. On older machines the clutch is disconnected using a large knob in the center of the handwheel, on newer machines the clutch is automatically disconnected with the bobbin winder is used.

Bobbin Winding Tension - The tension of the thread used during bobbin winding. If the tension is excessive it can damage the bobbin. If the tension is insufficient the thread may shift and cause problems during sewing.

Brother - A large Japanese based manufacturer of home and industrial sewing machines.

Button Hole Attachment - An attachment that is mounted in place of a presser foot and will allow a sewing machine to make button holes. This is used on machines that do not have a built in buttonholer.

Buttonholer - A function of some sewing machines that allows the machines to make button holes. The buttonholer function is usually controlled by knobs on the face of the machine.

C

Cam (stitch disk) - A cam for sewing machines is a disk used to make zigzag and decorative stitches. Some machines have built in cams while in other machines the cams are removable. Each cam disk makes one stitch type.

Cast Iron Body - Older home and industrial machines use cast iron metal for the body and bed of the machine. Cast iron is strong and rigid and the machines made of this metal are robust and have a long service life. However it is a very heavy material and has been replaced with aluminum (which is much lighter) in modern machines.

Class 15 Bobbin - This bobbin style originated with the Singer class 15 sewing machines but have become the most popular style of bobbin for home sewing machines and are now used by most manufacturers. There are two types, the original class 15 bobbin with straight sides that is available in both metal and plastic and the newer class 15J bobbin with tapered sides. Class 15J bobbins can be used in older machines. Original "Class 15" bobbins should not be used in newer machines that were designed for Class 15J bobbins.

Class 15 Sewing Machine - Singer introduced the model 15K in 1890 and went on to make many other class 15 models such as the 15-91, 15-88, etc. Later Japanese companies produced versions of these machine called HA-1 machines. All class 15 machines use a vertical oscillating hook.

Class 66 Bobbin - This bobbin style originated with the Singer class 66 sewing machines and were also used in many other Singer models such as the 99, 201, 400 series, 500 series and 600 series. Few other manufactures made models that used the class 66 bobbins and most newer machines use class 15 bobbins because they are larger and hold more thread.

Class 66 Sewing Machine - Used as a generic term "Class 66 sewing machine" can refer to any machine that uses a Class 66 bobbin including the 99, 201, 400 series, 500 series and most 600 series machines.

Coats & Clark - A manufacturer of thread based in the UK.

Computer Interface - A connection to a computer. Some electronic sewing machines or embroidery machines have a USB or Ethernet connection to connect the built in computer to an external computer such as a laptop or personal computer. This is used for transferring patterns or stitches to the machines computer.

Construction Seam - A seam that is used to assemble an item or for structural applications, in other words where stress could be applied to the seam during use. This is contrasted to a decorative seam that probably will not come under stress.

Construction Stitch - A stitch that is suitable for use in a construction seam.

Cursor Controls - Buttons on a sewing machine that control the position of the cursor on an LCD display.

Cutting Table - A table used for cutting fabric. If the table is to be used with a hand powered rotary cutter a cutting mat will be needed to cover the top of the table to prevent damage.

Cutter - One of several types of cutting tools for cutting fabric. There are electric rotary cutters, Hand rotary cutters, slide cutters and razor cutters.

D

Darning - Sewing to repair holes in fabric or reinforce fabric usually done by making a series of stitches back and forth over the area of concern and some of the surrounding fabric. Darning that is done with a sewing machine can be done using the forward and reverse operation of the machine but is more commonly done in free motion mode with the feed dogs disengaged.

Decorative Stitch - Any of many types of stitches used to primarily to decorate. Most decorative stitches are variations of the lockstitch but chain stitches can also be used for decoration such as in jeans using thick thread.

Decorative Stitch Cam - See Cam.

Default Settings (electronic machines) - Default settings are the settings that the machine powers on with or returns to after a reset. For example if the stitch width is set to 3mm and the needle position is set to the center, the machine may fall back to the default setting of 5mm stitch width and left needle position when it is restarted. Some machines will remember the last used setting when turned on and off, in this case the machine can said to power on "last set" instead of "with default settings".

Default Tension Setting - The recommended tension setting for general purpose thread (or in the case of heavy duty machines whatever thread the machine was designed for). The default tension setting may be indicated on the tension knob if it is numbered or could be stated in the owners manual. This should be the starting tension you use when you first start using the machine and is the setting that the machine was set up with at the factory.

Denim - A heavy twill fabric made of cotton used for jeans and outer wear.

Drive Belt - A belt that is made of rubber and fiber strands used to transfer power from the motor to the main shaft of a sewing machine. Drive belts come in many types including V-belt, cogged belt, notched belt, rubber round belt and leather belt.

Drive Belt Tension - The tightness of a drive belt in relation to the motor pulley and handwheel pulley. If the belt is too loose it will slip. If the belt is too tight it will cause accelerated wear.

Drive Belt Tension Adjustment - The adjustment to loosen or tighten drive belt tension, this is normally done by screws that allow the movement of the motor bracket on the frame of the machine.

Dropping Feed Dogs - Some machines are able to drop the feed dogs so that they do not contact the fabric at all while the machine is running. This is done to allow the machine to do free motion quilting and darning. With the feed dogs dropped the operator can move the fabric in any desired direction and control the way the stitches are formed.

Domestic Sewing Machine - Another name for home sewing machine.

E

Edging Stitch - Any stitch used for terminating the edge of fabrics. The most popular stitch for edging is the overlock stitch but the overlock stitch can not be made by a regular sewing machine (only a serger or an overlock machine) so there are also several overedge stitches that can be used. The overedge stitches are variations of zigzag stitches and can be made by a lockstitch sewing machine.

Edge Finishing - Applying an overlock stitch or overedge stitch to terminate (sew over to prevent fray) the edge of fabrics.

Edge Guide - A metal or plastic guide that can be attached to the bed (or sometimes the foot) of a sewing machine or serger to guide the fabric through the machine a precise distance from the needle.

Electronic - A device (sewing machine in this case) that uses electronic parts and circuits inside. Electronic circuits include parts such as computer chips, transistors, circuit boards, LCD displays, etc. Electronics enable the machine to have advanced features such as automatic thread cutters, automatic needle positioning, automatic needle up/down, automatic back tacking, fancy decorative stitch patterns, combinations of stitches, mirror imaging of stitches, monogramming, etc.

Electronic Sewing Machine - A sewing machine that uses electronics (see above). This is in contrast to a "mechanical" machine that does not use electronics.

Electronic Control - Control of sewing machine parts by electronic circuits and sensors. Some parts that could be electronically controlled include the motor, presser foot, needle bar position, thread cutter, etc.

Elna - A Swiss manufacturer of home sewing and embroidery machines. Also see the chapter Sewing Machine History.

Embroidery - The use of a needle and thread to decorate fabric. Embroidery can be done by hand or with a sewing or embroidery machine.

Embroidery Machine - A type of sewing machine designed for embroidery. Most embroidery machines are computer controlled. Some industrial embroidery machines have multiple heads and can sew several items at the same time.

Eye - An opening in a sewing machine needle just above the tip and below the scarf that the thread goes through.

F

Fabric Guide - See edge guide.

Feed Dogs - Moving metal parts with teeth use to pull fabric through a sewing machine. The feed dogs are mounted below the needle plate and their movement is controlled by a mechanism that causes them rise into contact with the fabric, move the fabric, lower out of contact with the fabric and then move to the forward position to start another cycle before rising again.

Feed Dog Cover Plate - A metal or plastic part used to cover the feed dogs so that a sewing machine can be used for free motion quilting or darning. With a feed dog cover plate in place the feed dogs do not control the movement of the fabric through the machine and the operator can move the fabric freely in any direction.

Feed Dog Height - The amount the feed dog teeth protrude through the needle plate when the feed dogs are in the up position. On some machines this can be controlled with a feed dog height control for best operation with thick and thin fabrics.

Feed Dog Height Control - See feed dog height.

Feed Dog Height Adjustment - See feed dog height.

Feed Dog Drop Lever - This is a control lever available on some machines used to drop the feed dogs for free motion quilting or darning. On machines without this lever a feed dog cover plate can serve the same purpose. Some machines have a feed dog height control that can also completely drop the feed dogs.

Feet - Plural of foot.

File Format - Computer files use different types of data such as text files, word files, graphics files, etc. These types of data are identified by file formats using what is called a file extension that comes after a period such as ".filetype" For example a text file named myfile would be "myfile.txt" and a graphic file of the jpeg type would be "mypicture.jpeg" Electronic sewing machines and embroidery machines use several file formats that are specific to embroidery data, one popular format is the .pes format. If you had a embroidery design in this format the file would be something like "myembroiderydesign.pes".

Flat Bed - The bed of a sewing machine is the area of the machine where the needle plate and feed dogs are located and the fabric passes over during sewing. In a flat bed machine this area is flat and broad allowing the fabric to be easily fed and controlled by the operator. Other types of beds are cylinder bed, post bed and free arms, these beds are used to sew odd shaped items with limited clearance such as sleeves.

Font - The design of lettering and numbering. Some popular computer fonts are Arial and Times New Roman. Some sewing machines and embroidery machines have many built in fonts.

Foot - The part of a sewing machine that holds the fabric in contact with the feed dogs to allow the fabric to move through the machine. There are many types of feet available for most machines.

Foot, Snap-On - A type of foot used in home sewing machines that snaps into a foot holder shank that mounts on the the presser bar of the sewing machine. Once the foot holder shank is mounted on the presser bar any snap-in foot can be snapped in or removed from the machine in seconds with no tools. Most modern home machines come with snap-in type feet.

Foot Controller - See foot pedal controller.

Foot Pedal Controller - A foot pedal with internal electric or electronic parts that is used to control the power output of a sewing machine motor. The foot pedal is connected to the motor with electrical wires. The motor gets no power when the pedal is not depressed. As your foot applies pressure and moves the pedal downwards electric power is made available to the motor. When the pedal is completely depressed the motor is at maximum power.

Free Arm - See cylinder bed.

Free Motion Quilting - Also see Dropping Feed Dogs. Free motion quilting is possible with most home sewing machines and some drop feed industrial sewing machines. The motion of the fabric through the machine is controlled by the operator and motion in any direction is possible. The operator can make artistic designs and patterns at will by controlling the speed of the machine and the movement of the fabric through the machine. Quite a bit of skill is required to do free motion quilting proficiently. There are specialized machines with long arms and beds that are specifically designed for free motion quilting and do not have feed dogs and therefore can not be used for normal sewing.

Front Loading Bobbin - This is a configuration of sewing machines where the bobbin is mounted vertically facing the front of the machine. This should not be confused with top loading machines in which the bobbin is mounted horizontally and drops in from the top.

G, H.

General Purpose Sewing Machine - Another name for the common lockstitch sewing machine. All home sewing machines and most industrial sewing machines could be classified as general purpose machines.

Gear Drive - The use of metal or plastic gears to interconnect or drive the motor or bottom and top shafts of sewing machines. Other means of driving these parts are belt drive or lever drive. Gear drive is characterized as being very reliable if metal gears are used, however plastic gears have a history of reliability problems due to aging and disintegration or cracking of the plastic over time.

Gutermann - A German manufacturer of thread.

HA-1 Bobbin Case - A very popular type of bobbin case used in class 15 sewing machines from many manufactures.

Hand Stitching Needle - A type of sewing needle made for hand sewing, the eye is located on the opposite end of the needle from the tip.

Handwheel - A flywheel and pulley mounted to the main shaft of a sewing machine with a smooth exposed outer surface to allow the rotation of the machine by hand when needed. Most machines have handwheels that rotate in the counterclockwise direction. The handwheel is used to slowly advance the machine for delicate stitching or to manually position the needle to the up or down position as needed. Also see Flywheel Effect.

Harp Space - The harp space of a sewing machine is the space between the needle and the body of the machine that is open for fabric to move during sewing. Most home machines have about 6-1/2 inches of harp space and most industrial machines have 9-/12 inches or more.

Head - The main frame and body of a sewing machine. This may or may not include the motor. Most modern home machines have the motor built in to the head of the machine and the head is the whole machine. Some older home machines are Treadle powered machines and the head is mounted to a table and is driven by a belt from the Treadle mounted in the bottom of the table. Most industrial machines have the head mounted in a heavy table and the motor is mounted below the table, the machine is driven by belt connecting the handwheel to the motor pulley.

Heavy Duty - For home machines a heavy duty machine is slightly heavier and more robustly constructed than a regular machine and can sew through somewhat thicker fabric or a few more layers of fabric. For industrial machines a heavy duty machine would be a very strong and tough machine with reinforced parts that can be used to sew leather, canvas, boat sails and other thick materials or many layers.

Hemming Machine - A specialized machine designed to a hemming stitch or blind stitch. These machines can not be used for general purpose sewing.

High Shank Foot - A type of foot mounting system used in some home sewing machines and most light and medium weight industrial drop feed sewing machines. Most standard high shank feet will work with any machine that accepts high shank feet.

Home Sewing Machine - A type of sewing machine specifically design for home sewing. Home machines are typically light weight and portable. They can be table mounted or mounted in cases. Most newer machines come with built in cases. Home machines are designed for intermittent or occasional operation and lack the forced oil lubrication systems of industrial machines.

Hook - A rotating or oscillating part of a lockstitch sewing machine that has a sharp point. When the needle is in the down position the point of the hook passes closely by the scarf of the needle and catches the needle thread forming a loop that the bobbin is then passed through to form the lockstitch.

Hook Mechanism - The hook and other mechanical parts of a sewing machine that drive the hook. The hook mechanism or hook assembly on a rotary hook machine usually consists of the hook, hook race, hook bearing and hook drive gear or belt. The hook mechanism on an oscillating hook machine usually consists of a shuttle-hook, race, race cover, shuttle driver and shuttle driver spring.

Hook Race - Part of the hook mechanism of a sewing machine that is used to contain the shuttle or bobbin case while it rotates or oscillates.

Hook-shuttle - See Shuttle-Hook.

Hoop - A circular or rectangular frame to hold the fabric in an embroidery machine. The hoop is comprised of two sections that snap together over the fabric or tighten together in some way to securely hold the fabric. After the fabric is inserted into the hoop the hoop is snapped or otherwise attached to the drive assembly of the machine and the machine can then control the motion of the fabric during sewing.

Horizontal Hook - A sewing machine hook that is horizontally mounted in the machine. Most machines with a horizontal hook also have a drop-in bobbin (also know as a top loading bobbin).

Horizontal Thread Spool - Used in some home sewing machines, also known as a Lay Down Thread Spool. The spool is mounted so that it is in a horizontal position. The thread comes off the top of the spool without the spool turning. This causes there to be less tension and smoother operation.

I, J, K

Imperial System - The Imperial System and the US Customary Measurement System are measurement systems consisting of inches, feet and yards used in the UK and USA. Most vintage sewing machines made in the UK and USA use parts that are dimensioned in these systems. Modern machines are all made using dimensions from the ISO Metric System. This is important to know because parts such as screws, nuts, shafts and bearings are not compatible between the two systems and can cause confusion if the user is not aware of this. A part can look like a perfect replacement yet not fit correctly because the systems are slightly different.

Industrial Sewing Machine - Industrial machines are designed for heavy use in manufacturing (factories). Sometimes they are also used in small business (commercial use) and by home sewers when extra durability or high speed is needed. Most industrial machines have forced oil lubrication systems and other features to allow for continuous used with decreased maintenance requirements.

Japanese Cast Iron Sewing Machine - Also see cast iron body. Almost all Japanese home sewing machines made before about 1965 were cast iron body machines with HA-1 style bobbin cases. There were millions of these machines made and they were sold under many brand names. These machines are of good quality and are reliable. There are straight stitch and zigzag models. They are very popular on the used market even today because of parts availability and reliability.

Japanese HA-1 Sewing Machine - See Japanese cast iron sewing machine above. Also see the chapter Sewing Machine History.

Janome - A large Japanese manufacturer of sewing machines, sergers and embroidery machines. Janome makes primarily home machines. Also see the chapter Sewing Machine History.

Juki - A large Japanese manufacturer of sewing machines that makes many models of Industrial sewing machines and overlock machines. Also see the chapter Sewing Machine History.

Kenmore - A US brand of sewing machines that is owned by the Sears chain of stores.

Knee Lifter - A lever operated by the users knee to raise the presser foot of a sewing machine. Most industrial machines have knee lifters. Some high end or heavy duty home machines have them.

Knife - see cutter.

L, M

Land - A slight bulge in a sewing machine needle between the eye and scarf. Some chain stitch needles have a second land between the scarf and the long groove to aid in the formation of the loop in machines that do not have needle bar rise.

LCD - Liquid Crystal Display. This is a type of display used in modern sewing machines and embroidery machines.

LED Light - Light Emitting Diode. A type of light that used very low power and has very long life. It is used for lighting and as indicator lights in sewing machines and other electronics.

Lock Point - The place where the upper and lower threads cross in a lockstitch. If the tension is adjusted correctly the lockpoint will be in the center of the fabric.

Lockstitch - A symmetrical type of stitch (the same on the top and bottom of the fabric). The lockstitch is made with a needle thread (needle thread) and a bottom thread (bobbin thread) that cross in the center of the fabric. The lockstitch is the most widely used stitch type for sewing machines and all home sewing machines and many industrial machines use the lockstitch.

Lock up - If a sewing machine is frozen and the handwheel cannot be turned it is said to be locked up.

Long Arm Quilting Machine - A type of sewing machine designed exclusively for free motion quilting. The long arm quilting machine does not have feed dogs and has a large harp space to handle large quilts. Long arm quilting machines are typically used with a quilting frame.

Lower Thread Path - The path of the bobbin thread in a lockstitch sewing machine or looper thread in a chainstitch machine or serger. This includes all guides, bobbin cases and tension assemblies.

Low Shank Foot - The type of foot mounting system used in most home sewing machines. Most standard low shank feet will work with any machine that accepts low shank feet.

Maximum Speed - The maximum speed that a sewing machine is rated to handle. This is usually given in SPM (stitches per minute). With home machines this is the maximum speed the machine will go with the built in motor. With industrial machines where the motors and machines can be purchased separately this is the maximum speed that the machine should be run at. The motor and pulley ratios should be selected to never exceed this speed or damage to the machine could occur.

Mechanical - Something that is physical as opposed to electronic. Gears, shafts and levers are all mechanical.

Mechanical Machine - A type of sewing machine that does not have electronics. All old vintage sewing machines are mechanical machines.

Memory - The ability and locations to save and recall settings on an electronic sewing machine. Memory is also used to store stitch patterns and embroidery designs.

Memory Recall - See above.

Metric System - The ISO (international) measurement system used on all modern sewing machines. Some vintage machines use the Imperial System. This is important to know because parts such as screws, nuts, shafts and bearings are not compatible between the two systems and can cause confusion if the user is not aware of this. A part can look like a perfect replacement yet not fit correctly because the systems are slightly different.

Mirror Image Stitch Function - A feature of some electronic sewing machines to invert a stitch (of fonts if you are sewing fonts). In this case the right side of the stitch will look like the left side and vise verse.

Mock Overlock Stitch - Same as overedge stitch. See overedge stitch.

Monogramming - The capacity of some sewing machines and embroidery machines to sew fonts (letters and numbers).

Motor - In the case of sewing machines all motors are electric motors. An electric motor converts electric power into a rotary mechanical motion of the output shaft. See Electric Motor. For home machines see Universal Motor. For industrial machines see Clutch Motor or Servo Motor.

Mounting Dimensions - Some home sewing machines are made for table mounting. There are two standard sizes for the openings in the table top and most machines will be one of the standard sizes, although there are a few brands of machines that use non standard dimensions. The small standard size opening is 7-1/8 inches x 14-3/4 inches and the large standard size opening is 7-1/8 inches x 16-1/2 inches.

N

Necchi - An Italian manufacturer of sewing machines. Also see the chapter Sewing Machine History.

Needle - See sewing machine needle.

Needle Bar - The needle bar in a sewing machine is a cylindrical metal part that is about the size of a pencil (or slightly smaller). The needle mounts to the needle bar with a needle clamp. The needle clamp holds the shank of the needle firmly to the needle bar. The needle bar is driven up and down by a linkage from the main shaft as the machine sews. The needle penetrates the fabric on the lower portion of its stroke. The needle bar is mounted in bearings that allow it to slide up and down easily as the machine operates.

Needle Deflection - As a sewing machine needle contacts and then penetrates the fabric during operation of the machine the needle will flex. This flex is known as needle deflection. A small amount of flex (under .25 mm) is normal and not a cause for concern but if there is a large amount of flex (more than . 5mm) the needle can collide with parts of the machine such as the needle plate of hook. This collision will cause the needle to permanently bend or break and could cause damage to the machine. Heavy fabrics require a larger needle size to minimize needle deflection, the larger needle diameter will make the needle stronger and stiffer. Excessive needle deflection and breakage can result from improper operation of the machine such as the operator pushing or pulling the fabric instead of letting the feed dogs do the work of moving the fabric though the machine.

Needle Hole - The small hole in the needle plate of a sewing machine that the needle goes through as it travels to its lowest position.

Needle Plate - The needle plate is a metal plate that covers the feed dogs and parts of the drive mechanism in the bed of a sewing machine. The needle plate has slots for the feed dogs to protrude through and a hole for the needle to go through. The needle plate attaches to the bed of the machine with screws or some sort of clamping parts and may be removed for cleaning and inspection of the feed dogs or to aid in visibility during lubrication and adjustment of the machine.

Needle Position - In a zigzag sewing machine there are three primary needle positions when the machine is used for straight stitching or reduced with zigzag stitches; left, center and right. Some machines have a control to select the needle position. In some machines there is a default position and no control to change to another position. Most machines default to the center position but some machines default to the left side. This is important to keep in mind because some feet and needle position combinations will not function causing the needle to hit the foot and break. Some electronic machines have continuously adjustable needle position controls and the needle can be precisely placed at any position. It is important to be cautious when powering on an electronic machine if you are using a special foot the requires a non-default needle position because some machines will not remember the last needle position used and return back to the default position. This could lead to needle breakage if the default needle position causes the needle to collide with with the special foot.

Needle Position Control - See needle position above.

Needle Shank - The top part of a sewing machine needle that is inserted into the needle clamp to hold the needle to the needle bar.

Needle Size - The size of the needle is determined by the diameter of the blade measured just above the scarf.

Needle System - Sewing machines needles have three major parameters, size, system and type. The term "Needle system" means "needle specification name".

Needle Thread - Also called upper thread, the needle thread is the thread that goes through the needle of a sewing machine.

Needle Threader - See automatic needle threader

Needle Type - Sewing machines needles have three major parameters, size, system and type. The type of a needle usually describes the tip and eye but may also pertain to other parts of the needle. Needles come in many types including sharp point, ball point, denim, twin, quilting, etc.

Needle Up/down Button - A button on an electronic sewing machine used to move the button to the up or down position.

Needle Up/down Control - A setting on an electronic sewing machine used to set the operation of the machine so that when the machine is stopped the needle will stop on the up position (normal) or the down position (for doing pivot turns). On some machines the needle up/down setting functionality is incorporated into the needle up/down button and there is not a separate control or menu setting.

O, P

Oil - See Sewing Machine Oil.

Oscillating Hook - The oscillating hook rotates back and forth (oscillates) about 200 degrees but never makes a full rotation. Also see Hook. Oscillating hooks are used in many home sewing machines. Sewing machines with an oscillating hook are limited to about 1600 SPM due to vibration that is caused by the oscillating motion.

Overcasting - See Edging Stitch.

Overlock Stitch - An overlock stitch is a multi-thread chain stitch in which at least one thread passes over the edge of the fabric. Overlock stitches are can be used for edge finishing or to join two pieces of fabric together and finish the edge in one operation. An overlock stitch can not be made by a lockstitch or general purpose sewing machine.

Owners Manual - A manual that is written by the manufacturer of a sewing machine for that specific model of machine. The owners manual will have exact information and instructions for a specific machine such as threading diagrams, lubrication requirements, tension adjustment procedures, etc.

Parts Manual - A manual that is written by the manufacturer of a sewing machine for that specific model of machine. The parts manual will list all parts used in the machine along with the part numbers. Most parts manuals also include detailed drawings and diagrams showing the location of the parts in the machine and what the parts look like.

Pattern disk - See Cam.

Pes File Format - A file format used for embroidery machines made by Brother and Baby Lock.

Pfaff - A German manufacturer of sewing machines, sergers and embroidery machines.

Pin Feed - A type of feed mechanism that uses a registration pin that protrudes form the feed dogs through the fabric to ensure that multiple layers of fabric will feed in synchronization without slippage. The pin is driven by the feed dog mechanism and pierces the fabric when the feed dogs rise, then retracts as the feed dogs drop after the fabric has been advanced.

Piecing Foot - A type of sewing machine foot often used in quilting for precisely joining small sections of fabric together.

Piercing power - The downward force that a sewing machine can apply to the needle bar to drive the needle through the fabric. Also see Flywheel Effect.

Plaid Matcher - Also called an Even Feed Foot. The Plaid matcher is an attachment foot used to help feed difficult fabrics evenly through the machine. The name "Plaid Matcher" is used because they are often used to help evenly feed plaid fabric so that pattern will match on seams. The plaid matcher is not a true walking foot like you find in an industrial machine. The attachment has two feet controlled by a mechanism that is powered by the needle bar of the machine. The feet slide with the movement of the fabric to reduce friction but do not actually push or power the fabric through the machine like a true walking foot.

Premium Sewing Machine Brand - Some brands of expensive home sewing machines are positioned in the market as premium brands. Premium brands sell lower quantities of machines at a higher profit. These machines usually have many features.

Presser Foot - The presser foot is the part of the sewing machine that presses down on the fabric from above and forces the fabric into contact with the feed dogs that are mounted in the bed of the machine below the needle plate. The sewing machine foot attaches to the shank and the shank attaches to the presser bar of the machine. Some feet are permanently attached to the shank while other feet use a snap-on style shank and are able to be removed from the shank. Also see Snap-On Foot.

Presser Foot Awareness - This is a feature of some electronic sewing machines. When a stitch is selected the machine will tell you (through the LCD screen) what foot to use for that particular stitch.

Presser Foot Lever - This is the hand lever mounted close to the presser bar that is used to raise and lower the presser foot. Some machines have both a knee lifter and a presser foot lever.

Presser Foot Lift - This is the clearance measurement between the bottom of the presser foot and the needle plate of a sewing machine when the presser foot is in the up position. Sometimes there are two levels that the foot can be raised to, the normal level and a high clearance level that can be achieved by further movement of the presser foot lever. On some machines that have knee lifters a higher level can only be achieved with the knee lifter and not the hand lever.

Presser Foot Pressure - The force that the presser foot presses down with when in the down (sewing) position. This force is adjustable on some machines by presser foot pressure control.

Programmable Stitches - Some electronic sewing machines have the capability to create, save and recall custom stitches or to modify, save and recall stitches.

Pulley - A special grooved wheel that is mounted to the output shaft of a motor or other rotating shaft. A belt is used between two pulleys to transfer rotational power from the driving pulley to the receiving pulley.

Q, R

Quilting Frame - A device to hold large quilts in a stationary manner and allow a free motion sewing machine to move on tracks and rollers. The machine can move to any part of the exposed quilt and sew in any direction. With a quilting frame and free motion sewing machine the user can sew complex patterns and designs. The quilting frame usually consists of two large rollers to hold the quilt that are suspended by a frame. The frame is mounted to a special heavy table for support. In addition there is a track system mounted to the table or frame that allows the machine to move freely on rollers.

Quilting Machine - See Long Arm Quilting Machine.

Raising the bobbin thread - On most mechanical lockstitch sewing machines you must raise the bobbin thread (bring the bobbin thread up through the needle hole) before you first use the machine and subsequently when you replace the bobbin after it runs out of thread. This is done by rotating the handwheel one full turn while holding the end of the needle thread with a slight bit of tension. As you rotate the handwheel the bobbin thread will be pulled up through the needle hole by the needle thread. Some electronic machines raise the bobbin thread automatically. Some machines with top loading bobbins do not need to have the bobbin thread raised to start sewing.

Rats Nest - This is a ball of tangled up thread that results from a malfunction in a sewing machine. The rats nest will usually occur in the hook and bobbin case area of the machine. If a rats nest occurs it must be carefully removed before the machine is used.

Reinforcement (at the beginning and end of seams) - Reinforcement is done at the beginning and end of seams by sewing about ½ inch (12mm) in the opposite direction (by putting the machine in reverse). This is done to prevent the stitches from pulling out at the beginning and end of the seam. Some sewing machines can do this automatically, they have a special button or can be programmed to do this at the start and end of every seam. In this case it is called automatic back tacking or automatic tacking.

Removable cam - See Cam.

Reverse Button, Spring Loaded - A type of reverse button or lever that causes the machine to go into reverse while the button or lever is depressed but automatically changes to forward operation when the button or lever is released.

Reverse Lever, Locking - A reverse lever or control that once set to reverse will stay in reverse until it is moved back to the forward position. This is handy when the machine must run in reverse for long seams because the operators hand can be used to guide the fabric instead of holding the lever.

Reverse Reinforcement Button - See Stitch Reinforcement.

Rotary Hook - A rotary hook makes full rotations and does not reverse direction (an oscillating hook makes a partial rotation and then reverses direction). Because a rotary hook rotates continuously in the same direction it results in smother operation then an oscillating hook and allows much higher speeds. Also see Hook. Rotary hooks are used in many models of home sewing machines and almost all industrial lockstitch machines.

S

Scarf - The scarf of a needle is the smooth flattened area of the shaft just above the eye of the needle. The scarf is flattened to allow the hook to pass between the shaft and the thread and make it easier for the hook to catch the thread.

Seam - A seam is a line of stitches sewn in succession. Seams that are used to attach or bind layers of fabric are known as construction seams. Seams can also be used for decoration.

Serger - Sergers are special machines that can only sew on the edge of the fabric. Sergers make a type of stitch called the overlock stitch. The overlock stitch is used for edge finishing. For more information about sergers check out our companion book "The Serger & Overlock Master Guide".

Serger Thread - General purpose serger thread is usually T27 and can be 2 or 3 ply. This is sold as serger thread in retail sewing stores.

Sewing Machine Needle - A sewing machine needle is a needle that is specifically designed for use in sewing machines and has the eye close to the point and a shank for attaching it to a sewing machine. The sewing machine needle also has a scarf to allow more clearance for the hook or looper as it passes by the needle.

Sewing Machine Oil - For sewing machines a clear or white "spindle oil" is used. This type of oil stains fabric as little as possible and has very low friction properties to reduce wear on the machine.

Shank (foot) - The part of a sewing machine foot that mounts to the presser bar of the machine. There are several types of shanks and they are not interchangeable. For home machines there are high shanks, low shanks, slant shanks and some manufactures use non-standard shanks. For industrial machines there are high shanks, Consew style walking foot shanks and several non-standard shanks.

Shank (needle) - See Needle Shank.

Shank Adapter - An adapter to allow a foot of one type to attach to a shank of a different type. For example one type of shank adapter allows mounting low shank feet on a high shank machine.

Shuttle-hook - The hook on an oscillating hook sewing machine part of a shuttle that slides in a race and also holds the bobbin-case. This single part that is a combination of shuttle and hook is called a shuttle-hook.

Slant Needle - See Slant Shank.

Shuttle Race - This is a metal containing ring and bearing that the shuttle-hook is located inside on an oscillating hook sewing machine.

Singer - A US manufacturer of sewing machines. Also see the chapter Sewing Machine History.

Slant Shank - A type of shank used on some Singer sewing machines such as the 400 series, 500 series, 600 series and 700 series. In a slant shank or slant needle machine the needle bar and needle are slanted towards the front of the machine so that the presser foot and feed dogs are about 1-1/4 inches closer to the front of the machine. This gives better visibility. It is a common myth that the slant needle design has superior piercing power but actually there is no advantage in piercing power because of the slant needle. These Singer models do have good piercing power but it is because of the spring coupled motor drive and not the slant needle.

Speed Range Control - This is a control on electronic machines that sets the top speed that the machine will run. On some home machines this control can also be used to control the operating speed of the machine if the foot pedal is not plugged in to the machine so that the machine can be operated with no foot pedal.

Spool Pin - A metal or plastic pin mounted to a sewing machine or thread stand to hold a spool of thread during sewing and prevent it from moving.

Stabilizer - A paper of synthetic material that is layered with fabric during sewing or embroidery to stop the fabric from stretching or otherwise moving while the machine is running. The stabilizer also helps the fabric to feed through the machine. The most stabilizer materials are designed to tear away after sewing.

Stitch - Interlocking, braided or chained threads used to attach, bind or decorate fabric or other material.

Stitch Cam or Stitch Disk - See Cam.

Stitch Length - The distance between stitches in a seam. Measured by counting the number of stitches per inch (SPI) or measuring the length of each stitch in millimeters. Newer machines use millimeters and some older machines use SPI. To convert stitch length in millimeters to SPI the equation is 25.4/mm=SPI. In this equation 25.4 is the number of millimeters in one inch. For example if you wanted to convert a 2mm stitch length to SPI then the equation would be 25.4/2=12.7 so your answer is 12.7 SPI.

Stitch Length Control - The length of each stitch the machine makes is controlled by the stitch length control.

Stitch Regulator - A type of control system used in some free motion quilting machines. The stitch regulator automatically controls the speed of the machine while the operator controls the movement of the fabric through the machine. The stitch regulator has a knob (or digital controls) to set the SPI (stitches per inch).

Stitch Reinforcement Button - This is a button that some electronic sewing machines have, pressing the button causes the machine to do an automatic back tack.

Stitch Width - The width of a zigzag or decorative stitch.

Stitch Width Control or Dial - A control on a zigzag sewing machine used to set the stitch width.

Stitches Per Inch (SPI) - A unit of measurement used to specify stitch length. SPI is not very intuitive for some people, because the higher the number of stitches per inch the shorter the stitch length. Also see Stitch Length.

Stitches Per Minute (SPM) - The speed of a sewing machine, the number of stitches the machine can sew in one minute.

Straight Stitch - A lockstitch or chainstitch in which the needle advances from stitch to stitch in a straight line to form a seam. This is in contrast to a zigzag stitch in which the needle moves from side to side.

Structural Stitch - A stitch that is used to assemble an item or for structural applications, in other words where stress could be applied to the seam during use. This is contrasted to a decorative stitch that probably will not come under stress.

T

Take-up Lever - The take-up lever is a lever in a sewing machine that removes the loop from the needle thread and tightens the stitch after the needle raises from the fabric.

Take-up Spring - The take up spring works with the take up lever to pull the thread gently as the loop is removed from the needle thread during the operation of a sewing machine. Without the take-up spring the take-up lever would yank the thread from the spool (causing the thread to fly around and possibly tangle up).

Tension - For the sewing machine to function and the stitches to be well formed the thread must be controlled and under the correct amount of tension or drag. Tension is a resistance or drag applied to the thread so that the thread will not freely unroll from the thread spool in an uncontrolled manner during sewing, leading to tangled thread and poor stitching.

Tension Adjustment Knob - A knob on a sewing machine used to control the tension of the needle thread (and looper threads in a chain stitch machine). Some machines have numbered tension adjustment knobs so that it is easy to remember an exact setting and return to that setting at a later date.

Tension Disks - To provide tension a sewing machine has adjustable friction disks or friction parts in the thread path and bobbin case. These friction disks or friction parts are pushed together by a spring and pinch or squeeze the thread to create tension.

Tension Screw - A screw on the side of a bobbin case that controls the bobbin thread tension. Turning the screw clockwise increases the tension and counterclockwise deduces the tension.

Tension Spring (on the thread tensioner assembly) - This is a coil spring that is positioned between the tension knob and the tension disks. This spring applies pressure to the tension disks.

Tension Spring (on the bobbin case) - A flat metal part that mounts to the bobbin-case with a screw (the tension screw). This plate is made of flexible metal (that is why it is called a spring). Tension is applied to the thread when is pinched between the tension spring and the bobbin case. Also see Tension Screw.

Tex System - The Tex system is the most widely used thread size measurement system and is used worldwide. It is the ISO standard (International Organization for Standardization). The Tex size is determined by the weight (in grams) for 1000 meters of thread. The Tex system is called a fixed length system because a fixed length of thread (1000 meters) is weighed to determine the thread size. The Tex system is intuitive because the numbers increase as the thread sizes get larger.

Thread Failure - Thread used in a seam or stitching breaking, rotting or become frayed.

Thread Guide - Any wire loop or other device on a sewing machine used for the purpose of controlling thread or routing thread along the thread path on its way through the machine

Thread Jam - See Rats Nest.

Thread Path - The path that thread takes when going though a threaded machine. The proper thread is given in the threading diagram for the machine or in the owners manual for the machine.

Thread Ply - A number of strands are twisted together to form most thread. These strands are also known as a ply, fold or yarn. For example a Three-ply thread is made from three strands or plies that have been twisted together.

Thread Ripper - A hand tool for removing seams from fabric.

Thread Twist - When thread is manufactured it is twisted to add strength and so that the strands will stay together. Most sewing machines are designed for thread with a left hand twist (Z twist) and most common thread is left hand twist. Thread with a right twist is known as "S twist". Thread of the correct twist must be used in sewing machines or the thread will snarl and knot while feeding through the machine.

Thread Size - There are many thread size systems and most of them are different and the sizes are not compatible between systems. The most popular thread size system is the TEX system. See the chapter Thread for more information on the many thread size systems.

Threading Diagram - A diagram that shows how to thread a sewing machine and the proper thread path of the machine. The threading diagram is part of the owners manual for the machine but may also be printed on the machine or the case of the machine. Most sergers have a threading diagram on the inside of the lower looper cover.

Thumb Screw - A machine screw with a large head that is designed to be turned by the fingers and thumb instead of requiring a screw driver. Thumb screws are used on most vintage sewing machines to attach the pressure foot.

Top Loading Bobbin - See Horizontal Hook.

Touch & Sew - A type of home sewing machine made by Singer in the 1970's. These models use something called a “Touch & Sew” or “wind-in-place” bobbin. Machines with the Touch & Sew bobbin have a track record of serious reliability problems although some users like them.

Twin Needle - This is a type of needle in which two needles are mounted to a single common shank. Twin needles are available with different spacing between the needles and are used to make two parallel rows of stitches at the same time.

U, V, W

Upper Thread Path - The path of the needle thread in a sewing machine or serger. This includes all guides, take-up lever, take-up spring and tension assemblies.

USB - Universal Serial Bus. This is a computer common interface that is used on some electronic sewing machines and embroidery machines to allow them to use USB memory sticks or be attached to a computer.

USB Interface - See USB.

Utility Stitch - A utility stitch is any lockstitch used for construction (joining fabric). The most common utility stitches are the straight stitch and the zigzag stitch but other stitches can be used as utility stitches such as the triple stitch. Most sewing is done with some form of utility stitch. See Construction Stitch.

V-Belt - A drive belt with tapered sides to match the tapered grove in a pulley. V-belts are used on most home sewing machines built before 1990 and most industrial sewing machines that are not direct drive. Also see Belt.

Variable Speed Controller - See Foot Controller.

Viking - A Swedish manufacturer of home sewing machines, embroidery machines and sergers. Also see the chapter Sewing Machine History.

Voltage - A unit of measurement of electricity. In the USA and some other countries 120V is used in homes. In Europe and many parts of Asia 240V is used. A sewing machine, electric light or any other type of equipment must be rated for (designed for) the correct voltage that is locally available, a 120V device can not be used with 240V electricity or damage and fire will result. The required voltage for most devices are listed on a label on the bottom or side of the device, sometimes close to the power cord. There are step down transformers and converters available to convert from 120V to 240V and visa verse.

Walking Foot - In a walking foot machine the presser is driven by the machine and foot moves with the feed dogs. When sewing multiple layers this will help to keep the top layers moving at the same rate as the bottom layers and prevents the top layers from sticking to the presser foot. The presser foot in these machines looks different than the presser foot in a drop feed machine and can have teeth like a feed dog. Some walking foot machines have a second smaller presser foot that holds the fabric stationary while the feed dogs and main presser foot recycle (move while out of contact with the fabric to get ready for the next stitch).

Walking Foot Attachment - See Plaid Matcher.

White - A US manufacturer of sewing machines.

X, Y, Z

Zigzag - The zigzag stitch is a lockstitch formed when the machine moves the needle from side to side evenly on every stitch (ISO stitch type 304).

Zigzag Width - The width of a zigzag stitch.

www.ingramcontent.com/pod-product-compliance
Lightning Source LLC
LaVergne TN
LVHW061204120826
845149LV00011B/1897